Navigation in Space by X-ray Pulsars

Amir Abbas Emadzadeh • Jason Lee Speyer

Navigation in Space by X-ray Pulsars

Amir Abbas Emadzadeh
University of California, Los Angeles
420 Westwood Plaza, 44-139-D1
Los Angeles, CA 90095-1597
USA
amire@ee.ucla.edu

Jason Lee Speyer
University of California, Los Angeles
420 Westwood Plaza, 38-1370
Los Angeles, CA 90095-1597
USA
speyer@seas.ucla.edu

ISBN 978-1-4419-8016-8 e-ISBN 978-1-4419-8017-5
DOI 10.1007/978-1-4419-8017-5
Springer New York Dordrecht Heidelberg London

Library of Congress Control Number: 2011921257

Printed on acid-free paper

Springer is part of Springer Science+Business Media (www.springer.com)

To my mom and dad, Batoul and MohammadKazem, my two sisters, Ensieh and Maryam, and Minoo, whom all I love so much (AAE)

To my wife Barbara, my children Gil, Gavriel, Rakhel, and Joseph, and my grandchildren Noa, Alon, Oliver, Isaac, Donovan, Vivian, and Sylvia for giving me so much love and joy (JLS)

Preface

This monograph, which is an extension of Dr. Emadzadeh's doctoral thesis, investigates the different aspects of utilization of X-ray pulsars for navigation of spacecraft in space. In our view, the monograph possesses two unique features. First, it provides a solid mathematical formulation for the absolute and relative navigation problems based on the use of X-ray pulsar measurements. Second, it presents a comprehensive framework for the signal processing techniques needed to obtain the navigation solution.

We had several readers in mind when writing this monograph. One such group is the body of university students and researchers who work on new space navigation techniques. Using X-ray pulsars for navigation is an interesting field in which there are many new challenging problems that need to be addressed. Another target group comprises people from the industry. Deep space navigation missions, especially the ones directed beyond the solar system, have attracted a lot of attention in recent years. Employing new space navigation techniques will definitely play a key role in making such missions successful. We hope that the monograph will encourage more researchers in the area of space navigation to work on X-ray pulsar-based navigation.

The subject matter requires some familiarity with linear systems, probability, and estimation theory. The knowledge is generally assumed to be of advanced undergraduate and graduate level. It will be more beneficial, if the reader proceeds through the chapters in sequence. We first provide some basic background knowledge on pulsars and a literature review on pulsar-based navigation in Chap. 2. Then, we present the navigation problems, and develop the X-ray pulsar signal models in Chap. 3. Using these models, we formulate and analyze the pulse delay estimation problem in Chap. 3. Different pulse delay estimators are proposed in Chaps. 4, 5, and 6. Using the presented estimators, Chap. 7 provides a recursive algorithm to obtain the navigation solution. Concluding remarks and suggestions for future work are given in Chap. 8.

Finally, we acknowledge Dr. A. Robert Golshan for his valuable comments and suggestions, which helped us greatly improve the monograph.

Los Angeles, CA
October 2010

Amir A. Emadzadeh
Jason L. Speyer

Contents

1 Introduction 1

2 Celestial-Based Navigation: An Overview 3
2.1 Introduction 3
2.2 Current Spacecraft Navigation Systems 3
2.3 Pulsar-Based Navigation 5
2.3.1 Why Celestial-Based Systems? 5
2.3.2 Pulsars 6
2.3.3 Why Use X-ray Pulsars for Navigation? 10
2.3.4 History of Pulsar-Based Navigation 10
2.4 Summary 12

3 Signal Modeling 13
3.1 Introduction 13
3.2 Proposed Navigation System Structure 13
3.3 X-ray Detectors 15
3.4 X-ray Pulsar Signal 16
3.4.1 Constant-Frequency Model 20
3.4.2 Time-Dependent-Frequency Model 20
3.5 Discussion 21
3.6 Epoch Folding 22
3.7 Epoch Folding in Presence of Velocity Errors 25
3.8 Generating Photon TOAs 28
3.9 Numerical Examples 30
3.10 Summary 33

4 Pulse Delay Estimation 35
4.1 Introduction 35
4.2 Pulse Delay Estimation 35
4.3 CRLB 36
4.4 Discussion 44
4.5 Numerical Examples 44
4.6 Summary 47

5 Pulse Delay Estimation Using Epoch Folding ... 49
5.1 Introduction ... 49
5.2 Cross Correlation Technique ... 49
5.3 Nonlinear Least Squares Technique ... 55
5.4 Discussion ... 57
5.5 Absolute Velocity Errors ... 58
5.6 Computational Complexity Analysis ... 59
5.6.1 Epoch Folding ... 59
5.6.2 CC Estimator ... 59
5.6.3 NLS Estimator ... 61
5.7 Numerical Examples ... 61
5.7.1 Initial Phase Estimators ... 61
5.7.2 Pulse Delay Estimators ... 64
5.8 Summary ... 70

6 Pulse Delay Estimation via Direct Use of TOAs ... 73
6.1 Introduction ... 73
6.2 Maximum-Likelihood Estimator ... 73
6.3 Numerical Determination of the MLE ... 77
6.4 ML Computational Complexity Analysis ... 78
6.5 Computational Complexity: Summary ... 78
6.6 Absolute Velocity Errors ... 79
6.7 Numerical Examples ... 79
6.7.1 ML Phase Estimator ... 80
6.7.2 Pulse Delay Estimator ... 81
6.7.3 Computational Cost ... 82
6.8 Summary ... 85

7 Recursive Estimation ... 87
7.1 Introduction ... 87
7.2 System Dynamics ... 87
7.3 Measurements ... 90
7.4 Discrete Time Estimation Process ... 92
7.5 Discussion ... 93
7.6 Absolute Navigation ... 94
7.7 A Geometric Approach for Estimation of Absolute Velocities ... 94
7.8 Numerical Examples ... 95
7.9 Summary ... 110

8 Epilog ... 111

References ... 113

Index ... 117

Acronyms

ARGOS	Advanced Research and Global Observation Satellite
CC	Cross correlation
CRLB	Cramér Rao lower bound
DARPA	Defense Advanced Research Projects Agency
DGPS	Differential Global Positioning System
DSN	Deep Space Network
EKF	Extended Kalman filter
GPS	Global Positioning System
IMU	Inertial measurement unit
JPL	Jet Propulsion Laboratory
ML	Maximum-likelihood
MLE	Maximum-likelihood estimator
MSE	Mean square error
NASA	National Aeronautics and Space Administration
NHPP	Non-homogeneous Poisson process
NLS	Nonlinear least squares
NRAO	National Radio Astronomy Observatory
NRL	Naval Research Laboratory
PSD	Power spectral density
RMS	Root mean square
SNR	Signal to noise ratio
SSB	Solar system barycenter
STD	Standard deviation
TOA	Time of arrival
VLBI	Very Long Baseline Interferometer
XNAV	X-ray Navigation

Chapter 1
Introduction

One of the main requirements for any space mission is to navigate the spacecraft. Current space navigation methods highly depend on ground-based operations. To achieve more autonomy and also to augment the current available navigation solutions, we are interested in utilization of celestial-based navigation systems. Such systems use signals emitted from celestial sources located at great distances from Earth. Because of their special characteristics, X-ray pulsars are potential celestial candidates for navigation.

The purpose of this book is to provide a concise treatment of utilizing a new navigation system for space missions, which is based on employing X-ray signals emitted from pulsars. Specifically, the book starts with reviewing current navigation methods being used for space missions. We provide motivations for adopting new space navigation approaches, and we suggest employing X-ray pulsars. We present the navigation system structure and provide the mathematical tools required to study and analyze different parts of the system. We also develop different signal processing techniques, which are essential to obtain the pulsar-based navigation solution.

Chapter 2 begins with an overview of current space navigation systems. It explains why utilizing celestial-based navigation systems is an interesting option for space missions. It shows why, of all available celestial sources, X-ray pulsars are suggested to be employed for space navigation. It also provides a brief treatment of pulsars and the history of pulsar-based space navigation.

In Chap. 3, the structure of the proposed X-ray pulsar-based navigation system is explained. Mathematical models describing the X-ray pulsar signals are developed. The time of arrival (TOA) of received photons is modeled as a non-homogeneous Poisson process (NHPP), and the probability density function of the TOAs is presented. Also, an effective algorithm is presented for simulation of the TOAs for any given pulsar with a known rate function. Additionally, it is explained how using *epoch folding*, photon intensity function can be retrieved by measurement of the TOAs. The noise associated with the procedure is also analyzed. Furthermore, the effect of imprecise spacecraft velocity information on epoch folding is studied.

Chapter 4 focuses on mathematical formulation of the pulse delay estimation problem. It employs models of the pulsar photon intensity functions on each detector. It also shows how to model the differential time between the spacecraft clocks.

A.A. Emadzadeh and J.L. Speyer, *Navigation in Space by X-ray Pulsars*,
DOI 10.1007/978-1-4419-8017-5_1, © Springer Science+Business Media, LLC 2011

Depending on available spacecraft data, the Cramér–Rao lower bound (CRLB) is provided for estimation of the unknown parameters. Some numerical examples are also presented.

In Chap. 5, it is proposed to first recover the pulsar photon intensity functions through the epoch folding procedure, and then estimate the pulse delay using the recovered photon intensities. Based on epoch folding, two different methods are proposed for the estimation of the pulse delay: (1) Using the cross correlation function between the empirical rate function and the true rate function and (2) Minimizing the difference between the empirical rate function and the true one through solving a nonlinear least squares problem. Asymptotic behavior of the proposed estimators and the effect of spacecraft velocity errors on their performance are also studied. Computational complexity of the estimators is investigated as well. Finally, analytical results are verified via numerical simulations.

Based on maximum-likelihood (ML) criterion and direct utilization of the measured photon time tags, another pulse delay estimator is proposed in Chap. 6. The estimator's asymptotic behavior, and the effect of imprecise spacecraft velocity data on its performance is studied. Computational complexity of the proposed estimator is investigated. And to assess the analytical results, numerical simulations are performed.

In Chap. 7, the dynamics models of inertial measurement unit (IMU) and spacecraft motion are employed by a Kalman filter to obtain the three-dimensional navigation solution. The effect of different navigation system parameters on achievable estimation accuracies is investigated. In particular, the effects of different values of IMU uncertainty, measurement noise variance, the number of pulsars used for measurement, their geometric dispersion in the sky map, and imprecise spacecraft velocity data are considered.

Finally, concluding remarks and suggestions for future work are provided in Chap. 8.

Chapter 2
Celestial-Based Navigation: An Overview

2.1 Introduction

In this chapter, we present an overview of spacecraft navigation using X-ray pulsars. In Sect. 2.2, we present a concise treatment of current navigation methods being utilized for space missions. Section 2.3.1 describes why employing celestial-based navigation techniques is desirable for space missions. We introduce different types of pulsars in Sect. 2.3.2. Section 2.3.3 explains why X-ray pulsars are interesting candidates to be used for navigation purposes. A short history of pulsar-based navigation is given in Sect. 2.3.4.

2.2 Current Spacecraft Navigation Systems

Most of space vehicle operations, thus far, have relied widely on Earth-based navigation methods for absolute position determination [1, 2]. Methods such as radar range and optical tracking are widely used for this purpose [3]. Although a ground-based tracking system has the advantage of not requiring an active hardware on the spacecraft itself, it does need extensive ground operations and careful analysis of the measured data in an electromagnetically noisy background environment. Also, as a spacecraft moves further away from Earth, its position estimation error increases if a radar-based navigation system is used. To achieve the necessary range determination, the radar system needs to know the observation station's position on Earth accurately. Another limitation is that such a system requires the knowledge of positional information of the solar system objects [1]. However, even if precise information of the radar station and solar system objects is available, the vehicle position estimation can only be accurate to a finite angular accuracy. The transmitted radar beam, along with the reflected signal, travels in a cone of uncertainty. This uncertainty degrades the position knowledge of the vehicle as a linear function of distance. Alternatively, many space vehicles, traveling into deep space or on interplanetary missions, employ active transmitters for orbit determination purposes [1]. The spacecraft receives a ping from an observation station on Earth

A.A. Emadzadeh and J.L. Speyer, *Navigation in Space by X-ray Pulsars*,
DOI 10.1007/978-1-4419-8017-5_2,

and retransmits the signal back to Earth. Then, the radial velocity is measured at the receiving station by measuring the Doppler frequency of the transmitted signal. Although some improvements are achieved in spacecraft navigation utilizing such systems, this method still has errors that increase with distance. Early experiments using these tracking systems on the Viking spacecraft showed accuracies to about 50 km in position estimation error for missions to Mars and positional accuracies on the order of hundreds of kilometers at the outer planets [2].

Another navigation approach is optical tracking. Spacecraft navigation based on optical tracking measurements is performed in a similar way as radar tracking [4]. This technique is based on the use of the visible light reflected from a vehicle to determine its location. For some optical measurements, a photograph needs to be taken and the vehicle's position is calculated after analysis of the photograph and comparison to a fixed star background. Therefore, real-time measurements using such systems typically are not achieved easily. Furthermore, optical measurements are limited by environmental conditions.

As many missions have concentrated on planetary observations, spacecraft navigation can be done by taking video images of the planet and comparing them to the known planetary parameters such as diameter and position relative to the other celestial objects. Throughout this procedure, the position of the spacecraft relative to the planet can be determined [5]. This requires the vehicle to be within the vicinity of the investigated planet.

To obtain accurate absolute navigation solutions for deep space missions, a combination of Earth-based radar ranging and on-vehicle planet imaging is typically required. This approach still requires human interaction and interpretation of data. Furthermore, as radar-ranging errors increase as the vehicle's distance from Earth increases, accurate navigation becomes more complex because of the required finer pointing accuracy of ground antennas. Additionally, the vehicles that process video images for navigation purposes need to have complicated onboard systems, which increase their cost. The imaging process also requires the vehicles to be sufficiently close to the planets. Therefore, it is necessary to investigate alternative methods that could provide a complete, accurate absolute navigation solution throughout the solar system, and perhaps eventually the intergalactic regimes.

For vehicles operating in space close to the Earth, the current Global Positioning System (GPS) can provide a complete autonomous navigation solution [6]. The GPS uses a constellation of between 24 and 32 medium Earth orbit satellites that transmit precise microwave signals, enable GPS receivers to determine their location, speed, direction, and time. However, these satellite systems have limited scope for the operation of vehicles relatively far from Earth.

For deep space missions, many spacecraft utilize the Deep Space Network (DSN). This system is an international network of antennas that supports interplanetary spacecraft missions, and radio and radar astronomy observations for the exploration of the solar system and the universe [7]. The network also supports selected Earth-orbiting missions. The DSN currently consists of three deep space communications facilities placed $\sim 120^{\circ}$ apart around the world: at Goldstone, in California's Mojave desert; near Madrid, Spain; and near Canberra, Australia.

This strategic placement permits constant observation of spacecraft as the Earth rotates and helps to make the DSN the largest and most sensitive scientific telecommunications system in the world.

Although accurate radial position can be determined using DSN, it requires extensive ground operations and scheduling to coordinate the observations. Even utilizing interferometry, the angular uncertainty still can increase significantly with distance. Position accuracies in the order of 1–10 km per astronomical unit (AU) of distance from Earth are achievable using interferometric measurements of the Very Long Baseline Interferometer (VLBI) through the DSN [1]. The VLBI is a type of astronomical interferometry used in radio astronomy. It allows observations of an object that are made simultaneously by many telescopes to be combined, emulating a telescope with a size equal to the maximum separation between the telescopes. Data received at each antenna in the array is paired with timing information, usually from a local atomic clock, and then stored for later analysis on magnetic tape or hard disk. At a later time, the data is correlated with data from other antennas similarly recorded to produce the resulting image. The resolution achievable using interferometry is proportional to the observing frequency and the distance between the antennas farthest apart in the array. The VLBI technique enables this distance to be much greater than that possible with conventional interferometry, which requires antennas to be physically connected by coaxial cable, waveguide, optical fiber, or other types of transmission line.

2.3 Pulsar-Based Navigation

2.3.1 Why Celestial-Based Systems?

Autonomous formation flying of multiple spacecraft is an important technology for both deep-space and near-Earth applications [8, 9]. One of the main requirements of a formation flight is accurate knowledge of the relative position and velocities between the vehicles. The spacecraft absolute navigation solution is also needed for any space mission. Several researchers have shown that the navigation solution for aerial and low-Earth-orbit applications can be obtained by utilizing differential GPS (DGPS). DGPS is an enhancement to Global Positioning System that uses a network of fixed, ground-based reference stations to broadcast the difference between the positions indicated by the satellite systems and known fixed positions. These stations broadcast the difference between measured satellite pseudoranges and actual (internally computed) pseudoranges, with receiver stations correcting their pseudoranges by the same amount. However, for deep space missions or situations where GPS is not available, an alternative approach is needed. Employing Earth-based navigation systems, such as DSN, is a possibility. But, as mentioned in Sect. 2.2, such systems suffer from low performance in situations where long range navigation is required. Furthermore, they are highly based on communicating with Earth to analyze their data.

Because of the aforementioned problems, the need for higher accuracy, and the continuing increase in cost of vehicle operations, spacecraft navigation is evolving from Earth-based solutions toward more autonomous methods [10, 11]. Autonomous operation means determination of a complete navigation solution by the spacecraft to guide itself toward its destination without human interaction or assistance. An autonomous navigation system internally computes its own navigation and guidance information by using onboard sensors. Any deviation from its planned path is detected, reported, and corrected without input from the ground mission control. These autonomous operations reduce the dependence of space missions on human interaction and communication with Earth. To reduce dependence of navigation systems on ground-based operations and achieve more autonomy, utilizing celestial-based navigation systems is desirable. Another reason for developing such novel navigation methods is to augment current systems by employing additional measurements to improve their performance. Celestial-based systems use signals emitted from celestial sources located at great distances from Earth. Of various celestial sources, X-ray pulsars are interesting candidates for use in both absolute and relative navigation systems because of their special characteristics. These characteristics are explained in detail in the following.

2.3.2 *Pulsars*

Celestial sources have played significant roles in navigation throughout history, although the majority of sources used have been fixed, persistent stars with visible radiation. Sources that produce signals with variable intensity are referred to as variable celestial sources. There are several classes of variable celestial objects emitting signals whose intensities vary from radio signals to gamma-ray over the electromagnetic spectrum. Of the different variable source types, ones producing a uniquely identifiable signal that is periodic and predictable can be utilized in a specific manner for navigation purposes. One particular class of variable celestial sources having this property is pulsars. Pulsars are rapidly rotating, highly magnetized neutron stars [12]. As the neutron star spins, charged particles are accelerated out along magnetic field lines in the magnetosphere. This acceleration causes the particle to emit electromagnetic radiation as a sequence of pulses produced and as the magnetic axis (and hence, the radiation beam) crosses the observer's line of sight in each rotation (see Fig. 2.1). The repetition period of the pulses is simply the rotation period of the neutron star. Pulsars are observed in the radio, visible, X-ray, and gamma-ray bands of the electromagnetic spectrum [13].

Radio pulsars are broadband, stellar pulsating radio sources powered by the rotation of a neutron star, resulting in a great stability in the pulsar period [14]. Over 1,300 pulsars are known [13], and more are being discovered through new research. In some pulsars, irregularities (glitches) in their rotational frequency are observed every few years, ranging from the order of 10^{-6} s for the Vela pulsar to only 10^{-8} s for the Crab pulsar. Although individually emitted pulses from the pulsars vary over

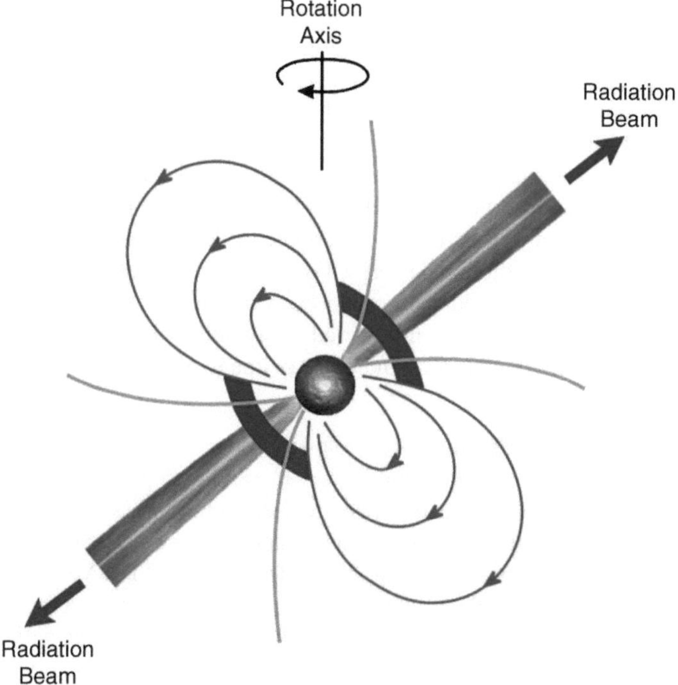

Fig. 2.1 Diagram of a pulsar. Photo courtesy of National Radio Astronomy Observatory (NRAO)

time, the average pulse shape is stable and characterizes the pulsar. Very precise models are established for the mean arrival time of pulsars, whose stability outperforms even the most precise artificial time bases. Of all pulsars, the most stable ones are the millisecond pulsars. Joseph Taylor and collaborators have demonstrated that the timing stability of millisecond pulsars is comparable with terrestrial atomic clocks [15]. Millisecond pulsars have been detected in the radio, X-ray, and gamma-ray portions of the electromagnetic spectrum. Currently, there are 130 millisecond pulsars known in globular clusters. Unfortunately, their signal to noise ratio (SNR) is considerably lower than that of longer period pulsars.

X-ray pulsars are grouped in two different categories according to the source of energy that powers the radiation: accretion-powered and rotation-powered pulsars [17].

1. Accretion-powered pulsars are a class of astronomical objects that are X-ray sources displaying strict periodic variations in X-ray intensity. The X-ray periods range from as little as a fraction of a second to as much as several minutes. An X-ray pulsar consists of a magnetized neutron star in orbit with a normal stellar companion and is a type of binary star system. The magnetic field strength at the surface of the neutron star is typically about 10^{12} Gauss, over a trillion times stronger than the strength of the magnetic field measured at the surface of the

Earth (0.6 Gauss) [18]. If the magnetic field and rotation axes of the neutron star are misaligned then X-ray pulsations are observed. Accretion pulsars are not stable timing sources because their period changes over time. More than 30 accretion-powered X-ray pulsars have been discovered with periods from 0.069 to 1,413 s [18].

2. Rotation-powered pulsars are rapidly rotating neutron stars whose electromagnetic radiation is observed in regularly spaced intervals, or pulses. They differ from other types of pulsars in that the source of power for the production of radiation is the loss of rotational energy. For a long time, the Crab pulsar, the most luminous rotation-powered pulsar, had been the only pulsar detected at X-ray energies. More than 20 rotation-powered X-ray pulsars have since been detected [18, 19]. Figure 2.2 depicts a Chandra X-ray image of the Vela rotation-powered pulsar (PSR B083345) [20].

Pulsars are the original gamma-ray celestial sources. A few years after the discovery of radio pulsars by astronomers, the Crab and Vela pulsars were detected at the gamma-ray band of the electromagnetic spectrum. Pulsars accelerate particles with tremendous energies in their magnetospheres. These particles are ultimately responsible for the gamma-ray emission seen from pulsars. The Vela pulsar, which spins 11 times a second, is the brightest persistent source of gamma rays in the sky. Yet gamma rays, the most energetic form of light, are few and far between. Even Fermi's Large Area Telescope sees only about one gamma-ray photon from Vela every 2 min. As opposed to a pulsar's radio beams which only content a few parts per million of its total power, gamma-rays represent 10% or more.

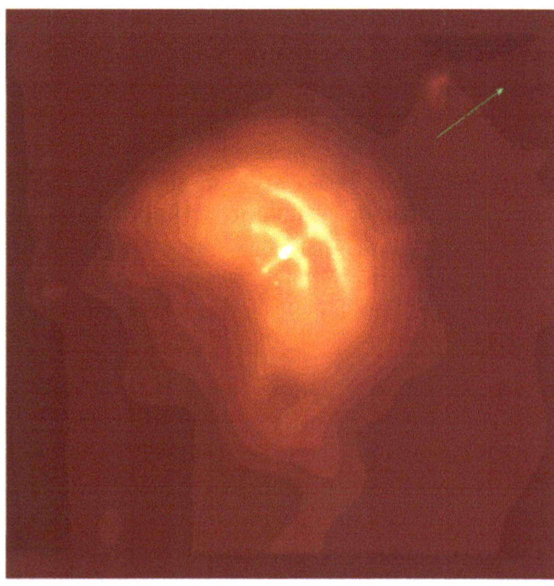

Fig. 2.2 Vela Pulsar (PSR B083345) X-ray image taken by Chandra X-ray observatory (Credit: The National Aeronautics and Space Administration (NASA)/PSU/G. Pavlov et al.)

By the end of 2004 there were about 1,500 radio pulsars known, but only seven had been detected in the gamma-ray band. Pulsars tend to have large magnetic fields and to be spinning rapidly. The loss of the pulsar's spin energy eventually appears as radiation across the electromagnetic spectrum, including gamma-rays. Both observations and models indicate that pulsars eventually lose the ability to emit gamma-rays as the pulsar's rotational speed slows down.

With NASA's Fermi gamma-ray space telescope, astronomers can now have a better look at pulsars. In two studies published in the July 2, 2009 edition of Science Express, international scientists have analyzed gamma-rays from two dozen pulsars, including 16 discovered by Fermi (see Fig. 2.3). Fermi is the first spacecraft able to identify pulsars by their gamma-ray emission alone [21]. The new pulsars were discovered as part of a comprehensive search for periodic gamma-ray fluctuations using 5 months of Fermi Large Area Telescope data and new computational techniques. In another part of the study, Fermi team examined gamma-rays from eight pulsars, all of which were previously discovered at radio wavelengths. Before Fermi launched, it was not clear that pulsars with millisecond periods could emit gamma rays. Now it is cleared that they do. It has also become clear that, despite their differences, both normal and millisecond pulsars share similar mechanisms for emitting gamma-rays.

NASA's Fermi Gamma-ray Space Telescope is an astrophysics and particle physics partnership, developed in collaboration with the U.S. Department of Energy, along with important contributions from academic institutions and partners in France, Germany, Italy, Japan, Sweden, and the U.S. [21].

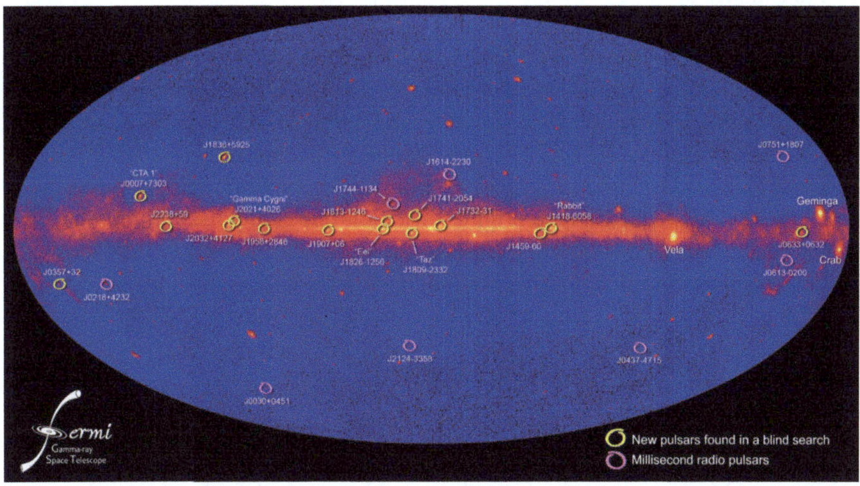

Fig. 2.3 This gamma-ray all-sky map, which is aligned with the plane of the Milky Way Galaxy, shows the pulsar positions, with the 16 new pulsars, detected by Fermi Gamma-ray Space Telescope, *circled* in *yellow* (eight previously known radio pulsars are in magenta) (Credit: NASA/DOE/Fermi LAT Collaboration)

2.3.3 Why Use X-ray Pulsars for Navigation?

Variable sources emitting signals in the radio band are certainly one potential candidate that can be used in a navigation system. However, at the radio frequencies that these sources emit, i.e., from 100 MHz to a few GHz, antennas with 20 m in diameter or larger are required to detect their signals [22,23]. At these wavelengths, "dish" style radio telescopes predominate. The angular resolution of a dish style antenna is a function of the diameter of the dish in proportion to the wavelength of the electromagnetic radiation being observed. This dictates the size of the dish that a radio telescope needs to have a useful resolution.

For most space missions, large antennas highly impact the design and cost of the operation [24]. Furthermore, because of neighboring sources that emit in radio bands and also low signal intensity of radio pulsars, long integration times are needed to obtain a signal with acceptable SNR, suitable for use in a navigation system. Similar limitations exist for the visible variable sources. Additionally, there are only five isolated pulsars known to emit in the visible band, and all are faint. There are also a only few pulsars discovered which emit in the gamma-ray wavelengths. This is another limitation for utilizing the visible and gamma-ray pulsars in a navigation system.

The disadvantages of the radio and visible sources diminish for sources that emit in X-ray band. The main advantage of spacecraft navigation using X-ray sources is that small sized detectors can be employed [25]. This provides savings in power and mass for spacecraft operations. Another advantage of using X-ray sources is that they are widely distributed. The geometric dispersion of pulsars in the sky is important to enhance accuracy of three-dimensional position estimation since the observability of the source is an important issue. An important complication that must be addressed in utilizing an X-ray source in a navigation system is the timing glitches in its rotation rates. Of X-ray pulsars, ones that are bright and have extremely stable and predictable rotation rates are suitable candidates for the purpose of navigation. These sources are usually older pulsars that have rotation periods on the order of several milliseconds. Figure 2.4 provides an image of the Crab Nebula and its pulsar (PSR B0531+21), which is the brightest rotation-powered pulsar within the X-ray band. The Crab pulsar shows high-flux X-ray emissions with known stable period. Hence, it can be considered a suitable candidate for use in navigation.

2.3.4 History of Pulsar-Based Navigation

The first pulsar was observed in July 1967 by Bell and Hewish. In 1971, Reichley, Downs, and Morris proposed using pulsar signals as a clock for Earth-based systems [26]. In 1980, details of methods to determine pulse time of arrivals from pulsar signals were provided by Downs and Reichley [27]. In the 1980s and 1990s, it was demonstrated that several pulsars matched the quality of atomic clocks [12, 15, 16, 28]. Because of their stability, pulsars were considered as accurate celestial clocks, suitable for navigation.

Fig. 2.4 Composite optical/X-ray image of the Crab Nebula pulsar showing surrounding nebular gasses stirred by the pulsar's magnetic field and radiation. Photo courtesy of NASA

In 1974, Downs, a member of the telecommunication division of the Jet Propulsion Laboratory (JPL), proposed a spacecraft navigation method based upon employing radio signals from a pulsar [29]. Using 27 radio pulsars for navigation over an integration time of 24 h, in [29], he showed that an absolute position accuracy on the order of 150 km was attainable. This introductory paper on pulsar navigation serves as the original basis for the work of other researchers in the field.

During the 1970s, pulsars with X-ray signature were discovered that emit signals within the X-ray band of 1–20 keV ($2.5e17 - 4.8e18$ Hz). In 1981, Chester and Butman proposed using X-ray pulsars as an option to enhance Earth satellite navigation [30]. Their research showed that by comparing the arrival times of pulses at a spacecraft and at the Earth (via an Earth orbiting satellite), a three-dimensional position of the spacecraft can be determined. They reported that a day's worth of data from a small onboard X-ray detector yielded a three-dimensional absolute position accurate to ~ 150 km.

In 1988, Wallace studied the issues related to using celestial sources that emit radio emission, including pulsars, for navigation applications on Earth [31]. He stated that the existence of other celestial radio sources obscured weak pulsar signals. As expected, radio-based systems require large antennas to detect sources, which make them impractical tools for spacecraft. Furthermore, the low signal intensity of radio sources requires long integration time to achieve an acceptable SNR. Also, the small population of radio pulsars in the optical band of the spectrum severely limits an optical pulsar-based navigation system.

In 1993, as a part of the NRL-801 experiment for the Advanced Research and Global Observation Satellite (ARGOS), Wood proposed a comprehensive approach to X-ray navigation covering attitude, position, and time. This study employed X-ray sources other than pulsars. As a part of the Naval Research Laboratory (NRL) development of this study, Hanson produced a Ph.D. thesis in the field of X-ray navigation in 1996 [32]. In his work, he studied attitude determination of spacecraft using X-ray pulsars. He used practical data from the HEAO-A1 spacecraft. His approach was based on counting the number of received photons, fitting the data to preknown curves and minimizing the Chi-squared error. He obtained roll estimates with error bias equal to 0.32 deg and standard deviation of 0.030° using a single detector and an error bias value equal to 0.012° with 0.0075° of standard deviation. He also suggested autonomous timekeeping using X-ray sources by employing a phase-locked-loop (PLL).

In 2004, a research group in Spain revealed a study on the feasibility of an absolute navigation system based on radio and X-ray pulsars [33]. The group developed some models of radio and X-ray pulsar signals, it proposed different algorithms for timing estimation of these two categories of celestial sources, studied their performance, and reported the possibility of obtaining absolute position accuracies on the order of 10^6 meters [33].

In 2005, Sheikh, a member of NRL, produced his Ph.D. thesis in the field of X-ray navigation [19, 34, 35]. His work was a part of a research called the X-ray Navigation (XNAV), which was directed by the Defense Advanced Research Projects Agency (DARPA). He proposed a navigation system based on X-ray measurements used by an extended Kalman filter (EKF) for three-dimensional position estimation [36, 37].

Woodfork suggested the use of X-ray pulsars for aiding GPS satellite orbit determination in his M.Sc. thesis in 2005 [38].

In 2009, Emadzadeh proposed a relative navigation algorithm based on use of X-ray pulsar measurements in his Ph.D. thesis [39]. He has studied different aspects of the signal processing techniques needed to obtain the X-ray pulsar measurements. His dissertation is the main reference of the current book.

2.4 Summary

This chapter presents a concise overview of current space navigation methods. It provides an introduction on different types of pulsars and their characteristics. It suggest using X-ray pulsars to navigate in space for situations where current systems are not available or it is desirable to augment the current navigation solutions. There are two main reasons that make X-ray pulsars interesting candidates for space navigation. One is their stable periodic profile, and the other is that relatively small size detectors are needed to detect the X-ray pulsar photons. The latter provides a huge advantage for the spacecraft design procedure. A brief history of pulsar-based navigation and previous research on this field is also discussed.

Chapter 3
Signal Modeling

3.1 Introduction

In this chapter, we present an overview of the X-ray pulsar-based navigation system. We provide the mathematical tools needed to analyze the proposed system. We also characterize the epoch folding procedure.

Section 3.2 describes the navigation problem and the proposed navigation solution. In Sect. 3.3, we explain how X-ray pulsar detectors work. Based on the presented X-ray detection mechanism, we use some mathematical tools to characterize the pulsar signals in Sect. 3.4. Section 3.5 offers a clarifying remark on pulsar signal modeling. We introduce and mathematically formulate the epoch folding procedure in Sect. 3.6. The effect of absolute velocity errors on epoch folding is studied in Sect. 3.7. An algorithm is provided in Sect. 3.8 for numerical simulation of the X-ray photon TOAs. To verify the analytical results, numerical examples are presented in Sect. 3.9.

3.2 Proposed Navigation System Structure

First, we consider the relative navigation problem between two spacecraft. At the end, we explain how the spacecraft absolute navigation problem can be addressed.

For relative position estimation, both the space vehicles must lock on the same X-ray source, and detect the same signal emitted from it. The spacecraft located farther away from the pulsar, receives a delayed version of the signal which is observed by the spacecraft closer to the source. The distance between the space vehicles is proportional to the time delay. The proposed approach is to periodically estimate the time delay, and then use these estimates as measurements in a recursive algorithm for position estimation. As it can be seen in Fig. 3.1, the relative distance projected on the direction of the source, Δd, is related to the time delay, t_{d}, by

$$\Delta d = \mathbf{n} \cdot \Delta \mathbf{x} = c t_{\mathrm{d}}, \tag{3.1}$$

where c is the speed of light, $\Delta \mathbf{x}$ the relative position vector, and $\mathbf{n}$ is the normalized direction vector pointing to the source.

A.A. Emadzadeh and J.L. Speyer, *Navigation in Space by X-ray Pulsars*,
DOI 10.1007/978-1-4419-8017-5_3, © Springer Science+Business Media, LLC 2011

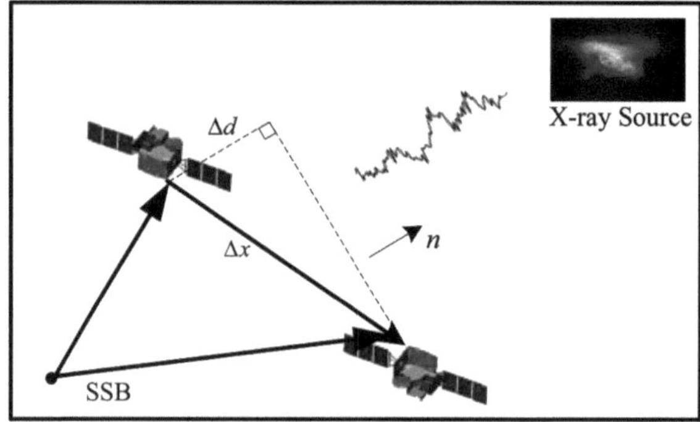

Fig. 3.1 Relative position between two spacecraft observing the same X-ray source

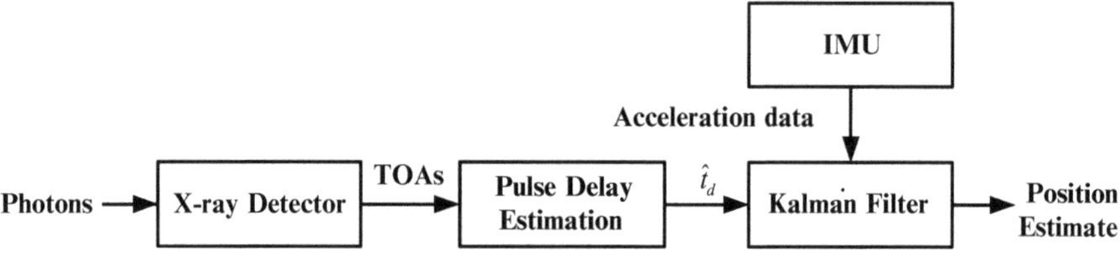

Fig. 3.2 Structure of the proposed X-ray pulsar-based navigation system

The pulsar is far away compared to the relative separation of the spacecraft, and the vehicles are assumed to be close enough. Hence, they see the source with the same direction vector, and this vector is known. All the measurements are performed in an inertial frame whose origin is the solar system barycenter (SSB). The spacecraft must communicate constantly to share and analyze their data. This communication channel can be used as an additional information source that can be incorporated to enhance the navigation solution. For simplicity, it is assumed that the cycle ambiguity problem does not arise. In other words, the signal passes the projected distance between the spacecraft, Δd, in less than one pulsar cycle. Relativistic effects are negligible. The measurement noise variance is selected based on the accuracy of the time delay estimates. On each spacecraft, an inertial measurement unit (IMU) is to provide the acceleration measurements, which are converted to the velocity and position data and are utilized by the Kalman filter. The time delay estimates are in turn taken in as measurements by the Kalman filter to obtain the navigation solution. The precision of relative range is based on the geometric distribution of the set of employed pulsars over the sky, in much the same way as the geometric distribution of GPS satellites determines the geometric dilution of precision. The proposed navigation algorithm can be summarized into the following stages (see Fig. 3.2).

1. Signal detection using X-ray detectors (Chap. 3).
2. Pulse time delay estimation (Chaps. 4, 5, and 6).
3. Obtain the navigation solution utilizing a Kalman filter (Chap. 7).

Note that if the first spacecraft is at the SSB, the navigation problem reduces to the absolute navigation problem of one spacecraft.

3.3 X-ray Detectors

X-ray detectors are designed based on measuring the time of arrival (TOA) of photons when they hit the detecting material [33]. To obtain the X-ay pulsar measurements precisely, a low power detection system is required that is capable of measuring the photon time of arrivals with sub-microsecond accuracy. Detectors must have a large detection area, low background noise, and be lightweight. Furthermore, they need to not require heavy and expensive cooling systems. Previous satellite missions have used X-ray detectors with gas-based technology. These detectors are prone to leaks in the entrance window [25]. Another disadvantage of the detector is that, although made of low density gas, they can be rather heavy because of the massive frames they need to contain the gas. Because of these problems, an alternative technology is much of interest to make X-ray detectors.

As part of DARPA XNAV program, NRL has developed a new technology for the next generation of X-ray silicon-based detectors that achieve all of the aforementioned capabilities [25]. These detectors are made of silicon PIN (p-type, intrinsic, and n-type) diodes that are reverse-biased with high voltage so that the whole silicon volume does not have any free charge carrier. Therefore, when an X-ray photon hits the silicon, it will create a large number of electrons and holes. These electrons and holes drift under the electric field to the surface. Then, they are collected into preamplifiers and generate signals whose amplitudes are proportional to the energy deposited by the X-ray photons. These signals are processed by a shaping amplifier and a discriminator to provide the time of arrival of photons. The time resolution of the X-ray detector is dictated by the drift time of the electrons across the thickness of the detector. The drift time in silicon at room temperature is 20 ns/mm [25]. The thickness of the detectors determines if they can efficiently detect higher energy X-rays. To obtain more accurate solutions for X-ray-based navigation, many X-ray photons must be collected. Therefore, the energy range of detectors must be maximized, meaning they must be made as thick as possible. The thickest silicon detectors currently manufactured are 2 mm thick [25]. More details of the silicon-based technology, including required hardware for post-processing of data, are explained in [25].

Based on this photon detection approach, mathematical models describing the X-ray pulsar signals are provided in the next section.

3.4 X-ray Pulsar Signal

Let (t_0, t_f) be the observation interval and $T_{\text{obs}} = t_f - t_0$. Furthermore, let t_i be the TOA of the ith photon and the set $\{t_1, t_2, \ldots, t_M\}$, denoted by $\{t_i\}_{i=1}^{M}$, be a random sequence in increasing order,

$$t_0 \leq t_1 < t_2 < \cdots < t_M \leq t_f, \tag{3.2}$$

where the number of detected photons in (t_0, t_f), M, is also a random variable.

Let $t_0 = 0$ and $N_0 = 0$ then

$$N_t = \max\{n,\ t_n \leq t\} \tag{3.3}$$

is called a *point process*, denoted by $\{N_t,\ t > 0\}$. Note that N_t is the number of detected photons in the interval $(0, t)$. A point process $\{N_t,\ t > 0\}$ is called a non-homogeneous Poisson process (NHPP) with a time-varying rate $\lambda(t) \geq 0$ if it satisfies the following conditions [40].

1. The probability of detecting one photon in a time interval Δt, is given by

$$P(N_{t+\Delta t} - N_t = 1) = \lambda(t)\Delta t \quad \text{when} \quad \Delta t \to 0. \tag{3.4}$$

2. The probability of detecting more than one photon in Δt is

$$P(N_{t+\Delta t} - N_t \geq 2) = 0 \quad \text{when} \quad \Delta t \to 0. \tag{3.5}$$

3. Non-overlapping increments are independent, where

$$N_t - N_s, \quad t > s \tag{3.6}$$

is the increment of the stochastic process $\{N_t, t > 0\}$.

For a fixed t, the number of detected photons in $(0, t)$, N_t, is a Poisson random variable with parameter $\int_0^t \lambda(\xi)\mathrm{d}\xi$ [40], that is

$$P(N_t = k) = \frac{\left(\int_0^t \lambda(\xi)\mathrm{d}\xi\right)^k \exp\left(-\int_0^t \lambda(\xi)\mathrm{d}\xi\right)}{k!} \tag{3.7}$$

and its mean and variance are

$$\begin{aligned} E[N_t] &= \operatorname{var}[N_t] \\ &= \int_0^t \lambda(\xi)\mathrm{d}\xi \\ &\triangleq \Lambda(t). \end{aligned} \tag{3.8}$$

The number of detected X-ray photons in any fixed time interval (t,s), $N_t - N_s$, is a Poisson random variable as well with parameter $\int_s^t \lambda(\xi)\mathrm{d}\xi$ [40], i.e.,

$$P(N_t - N_s = k) = \frac{\left(\int_s^t \lambda(\xi)\mathrm{d}\xi\right)^k \exp\left(-\int_s^t \lambda(\xi)\mathrm{d}\xi\right)}{k!}. \tag{3.9}$$

The probability density function of the photon TOAs is presented by the following theorem [41].

Theorem 3.1. *The M-dimensional joint probability density function (pdf) of the TOAs set, $\{t_i\}_{i=1}^M$, denoted by $p\left(\{t_i\}_{i=1}^M, M\right)$, is given by*

$$p\left(\{t_i\}_{i=1}^M, M\right) = \begin{cases} \mathrm{e}^{-\Lambda} \prod_{i=1}^{M} \lambda(t_i) & M \geq 1 \\ \mathrm{e}^{-\Lambda} & M = 0, \end{cases} \tag{3.10}$$

where

$$\Lambda \triangleq \Lambda(t_f) - \Lambda(t_0) \tag{3.11}$$

is called the integrated rate *of the Poisson process.*

Proof. To calculate $p(\{t_i\}_{i=1}^M, M)$, consider non-overlapping infinitesimal intervals of width Δt_i symmetric about t_i, $i = 1, 2, \ldots, M$ (see Fig. 3.3). The TOAs pdf, satisfies the following,

$$\begin{aligned} &P\left[\tau_1 \in \left(t_1 - \frac{\Delta t_1}{2}, t_1 + \frac{\Delta t_1}{2}\right), \ldots, \tau_M \in \left(t_M - \frac{\Delta t_M}{2}, t_M + \frac{\Delta t_M}{2}\right)\right] \\ &= \int_{t_1 - \Delta t_1/2}^{t_1 + \Delta t_1/2} \cdots \int_{t_M - \Delta t_M/2}^{t_M + \Delta t_M/2} p(\{\tau_i\}_{i=1}^M, M)\mathrm{d}\tau_1 \cdots \mathrm{d}\tau_M. \end{aligned} \tag{3.12}$$

The expression on the left-hand side of (3.12) is the probability of detecting exactly one photon in each one of the intervals and none outside them. In other words,

$$\begin{aligned} &P\left[\tau_1 \in \left(t_1 - \frac{\Delta t_1}{2}, t_1 + \frac{\Delta t_1}{2}\right), \ldots, \tau_M \in \left(t_M - \frac{\Delta t_M}{2}, t_M + \frac{\Delta t_M}{2}\right)\right] \\ &= P(N_{t_1 - \Delta t_1/2} - N_{t_0} = 0) \cdot P(N_{t_1 + \Delta t_1/2} - N_{t_1 - \Delta t_1/2} = 1) \cdot \\ &\qquad \vdots \\ &P(N_{t_M + \Delta t_M/2} - N_{t_M - \Delta t_M/2} = 1) \cdot P(N_{t_f} - N_{t_M + \Delta t_M/2} = 0). \end{aligned} \tag{3.13}$$

Fig. 3.3 A sample realization of a Poisson point process

Using (3.9), the probability of receiving $M = 0$ photon in the interval (a,b) is given by

$$P(N_b - N_a = 0) = \exp\left(-\int_a^b \lambda(t)\mathrm{d}t\right). \tag{3.14}$$

Therefore, by letting $\Delta t_i \to 0$ for all i, and substituting (3.4) and (3.14) into the right-hand side of (3.13), it can be simplified as

$$P\left[\tau_1 \in \left(t_1 - \frac{\Delta t_1}{2}, t_1 + \frac{\Delta t_1}{2}\right), \ldots, \tau_M \in \left(t_M - \frac{\Delta t_M}{2}, t_M + \frac{\Delta t_M}{2}\right)\right]$$
$$= \mathrm{e}^{-\Lambda} \prod_{i=1}^{M} \lambda(t_i)\Delta t_i, \tag{3.15}$$

where Λ is given in (3.11).

On the other hand, as $\Delta t_i \to 0$ for all i, right side of (3.12) equals

$$\int_{t_1-\Delta t_1/2}^{t_1+\Delta t_1/2} \cdots \int_{t_M-\Delta t_M/2}^{t_M+\Delta t_M/2} p\left(\{\tau_i\}_{i=1}^M, M\right) \mathrm{d}\tau_1 \cdots \mathrm{d}\tau_M = p\left(\{t_i\}_{i=1}^M, M\right) \prod_{i=1}^{M} \Delta t_i. \tag{3.16}$$

Therefore, since (3.15) and (3.16) are equal, the joint pdf $p\left(\{t_i\}_{i=1}^M, M\right)$ is obtained for $M \geq 1$,

$$p\left(\{t_i\}_{i=1}^M, M\right) = \mathrm{e}^{-\Lambda} \prod_{i=1}^{M} \lambda(t_i). \tag{3.17}$$

■

The overall rate function $\lambda(t) \geq 0$, represents the aggregate rate of all arriving photons from the X-ray pulsar and background,

$$\lambda(t) = \lambda_\mathrm{b} + \lambda_\mathrm{s} h\big(\phi_\mathrm{det}(t)\big) \quad \text{(ph/s)}, \tag{3.18}$$

where $h(\phi)$ is the periodic *pulsar profile*, $\phi_\mathrm{det}(t)$ the *detected phase*, and, λ_b and λ_s are the known *effective background* and *source arrival rates*, respectively [42]. The periodic function $h(\phi)$ is usually defined on the phase interval $\phi \in [0,1)$ (cycle), and then its definition is extended to the entire real line by letting $h(\phi+n) = h(\phi)$, where n is an integer. In other words, each period of the pulsar intensity is represented by one cycle. Furthermore, the function $h(\phi)$ is non-negative and normalized according to $\int_0^1 h(\phi)\mathrm{d}\phi = 1$, and $\min_\phi\ h(\phi) = 0$. The shape of the pulsar profile and its period are known. Figure 3.4 shows the profile of the Crab pulsar (PSR B0531+21) in the X-ray band (1–15 keV), created using multiple observations with the USA experiment onboard the Advanced Research and Global Observation Satellite (ARGOS) vehicle [19].

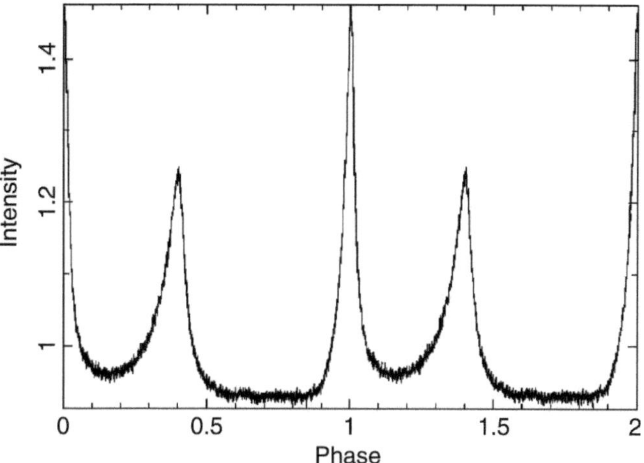

Fig. 3.4 Crab pulsar profile obtained by ARGOS

The detected rate constants λ_b and λ_s are functions of the detector specifications. Specifically, they are proportional to the collective area, A, and the detector efficiency, η, as

$$\lambda_b = b \cdot \eta \cdot A \tag{3.19}$$

$$\lambda_s = s \cdot \eta \cdot A, \tag{3.20}$$

where b is the background rate and s is the source flux.

Phase at the detector referenced to the beginning of the observation, t_0, consists of an initial phase, $\phi_0 \in [0,1)$, and the accumulated phase

$$\phi_{\text{det}}(t) = \phi_0 + \int_{t_0}^{t} f(\tau)\mathrm{d}\tau, \tag{3.21}$$

where $f_o(t)$ is the observed signal frequency, which can be decomposed into two different components: X-ray source frequency, f_s, and the Doppler frequency shift, $f_d(t)$. In other words,

$$f_o(t) = f_s + f_d(t). \tag{3.22}$$

The Doppler frequency $f_d(t)$ is due to the detector's velocity, $v(t)$, and is given by

$$f_d(t) = f_s \frac{v(t)}{c}, \tag{3.23}$$

where c is the speed of light. Consequently, the detected phase equals

$$\phi_{\text{det}}(t) = \phi_0 + f_s(t - t_0) + \int_{t_0}^{t} f_d(\tau)\mathrm{d}\tau. \tag{3.24}$$

Note that the pulsar is located far deep in space (at infinite distance from the detector). Furthermore, the observed phase is calculated referenced to the beginning of the observation time, and not the time that photons have left the X-ray source. Hence, although $h(\cdot)$ is periodic, the phase ambiguity issue does not arise.

3.4.1 Constant-Frequency Model

Assuming that the detector's velocity $v(t) = v$ is a known constant, then

$$\phi_{\text{det}}(t) = \phi_0 + (t - t_0) f_o, \tag{3.25}$$

where

$$f_o = \left(1 + \frac{v}{c}\right) f_s. \tag{3.26}$$

Hence, from (3.18), the rate function is

$$\lambda(t; \phi_0, f_o) = \lambda_b + \lambda_s h\big(\phi_0 + (t - t_0) f_o\big). \tag{3.27}$$

This model is used when the spacecraft moves in a constant radial speed and the observed pulse frequency, f_o, does not change over time or its change is negligible. Since $h(\cdot)$ is strictly periodic and its argument is an affine function of time, $\lambda(t)$ becomes strictly periodic as well.

3.4.2 Time-Dependent-Frequency Model

If $v(t)$ is not constant, $f_d(t)$ is a time dependent, and

$$\phi_d(t) = \int_{t_0}^{t} f_d(\tau) \mathrm{d}\tau \tag{3.28}$$

is a nonlinear function of time, which results in a quasi-periodic $\lambda(t)$. This is the case when the radial speed of the spacecraft, $v(t)$, changes significantly over the observation time. Hence, the observed phase at the detector is

$$\phi_{\text{det}}(t) = \phi_0 + (t - t_0) f_s + \phi_d(t) \tag{3.29}$$

and the rate function becomes

$$\lambda\big(t; \phi_0, v(t)\big) = \lambda_b + \lambda_s h\big(\phi_0 + (t - t_0) f_s + \phi_d(t)\big). \tag{3.30}$$

In (3.30), presence of the nonlinear term, $\phi_d(t)$, makes the Poisson rate function be quasi-periodic.

3.5 Discussion

Using the TOA pdf presented in (3.10), it can be verified that

$$\int_{\Omega} p\left(\{\tau_i\}_{i=1}^{M}, M\right) \mathrm{d}\tau_1 \cdots \mathrm{d}\tau_M = 1, \tag{3.31}$$

where Ω is the sure event, i.e., the event that any number of photons occur at any time instant in the interval $[t_0, t_f]$. Note

$$\int_{\Omega} p\left(\{\tau_i\}_{i=1}^{M}, M\right) \mathrm{d}\tau_1 \cdots \mathrm{d}\tau_M = \sum_{M=0}^{\infty} \int_{\Omega_M} p\left(\{\tau_i\}_{i=1}^{M}, M\right) \mathrm{d}\tau_1 \cdots \mathrm{d}\tau_M, \tag{3.32}$$

where Ω_M is the event of receiving M photons at any M different increasing time instants $\{t_1, t_2, \ldots, t_M\}$ in the interval $[t_0, t_f]$. The probability of event Ω_M is given by

$$\begin{aligned} &\int_{\Omega_M} p\left(\{\tau_i\}_{i=1}^{M}, M\right) \mathrm{d}\tau_1 \ldots \mathrm{d}\tau_M \\ &\quad = P[\mathbf{t}_1 = t_1 \in (t_0, t_f), \mathbf{t}_2 = t_2 \in (t_1, t_f), \ldots, \mathbf{t}_M = t_M \in (t_{M-1}, t_f)]. \end{aligned} \tag{3.33}$$

Note that the sequence $\{t_1, t_2, \ldots, t_M\}$ in (3.33) is in increasing order. Since there exists $M!$ permutations of t_i, for a fixed M, the probability that M number of t_i's occur in no special order, is $M!$ times that of the sequence occurring in increasing order [41]. Hence,

$$\begin{aligned} &P[\mathbf{t}_1 = t_1 \in (t_0, t_f), \mathbf{t}_2 = t_2 \in (t_0, t_f), , \ldots, \mathbf{t}_M = t_M \in (t_0, t_f)] \\ &\quad = M! \cdot \int_{\Omega_M} p\left(\{\tau_i\}_{i=1}^{M}, M\right) \mathrm{d}\tau_1 \cdots \mathrm{d}\tau_M. \end{aligned} \tag{3.34}$$

Using (3.10), the left side of (3.34) can be expressed as

$$\begin{aligned} \int_{t_0}^{t_f} \int_{t_0}^{t_f} \cdots \int_{t_0}^{t_f} p\left(\{\tau_i\}_{i=1}^{M}, M\right) \mathrm{d}\tau_1 \cdots \mathrm{d}\tau_M &= \mathrm{e}^{-\Lambda} \left(\int_{t_0}^{t_f} \lambda(\tau) \mathrm{d}\tau \right)^M \\ &= \mathrm{e}^{-\Lambda} \Lambda^M. \end{aligned} \tag{3.35}$$

Therefore, from (3.34) and (3.35), for a fixed M

$$\int_{\Omega_M} p\left(\{\tau_i\}_{i=1}^{M}, M\right) \mathrm{d}\tau_1 \cdots \mathrm{d}\tau_M = \frac{\Lambda^M}{M!} \cdot \mathrm{e}^{-\Lambda}. \tag{3.36}$$

Now, using (3.32) and (3.36)

$$\begin{aligned}\int_{\Omega} p\left(\{\tau_i\}_{i=1}^{M}, M\right) \mathrm{d}\tau_1 \cdots \mathrm{d}\tau_M &= \mathrm{e}^{-\Lambda} \sum_{M=0}^{\infty} \frac{\Lambda^M}{M!} \\ &= \mathrm{e}^{-\Lambda} \cdot \mathrm{e}^{\Lambda} \\ &= 1. \end{aligned} \tag{3.37}$$

3.6 Epoch Folding

A feasible question is how to recover the pulsar rate function employing the measured photon TOAs. The procedure of recovering the X-ray pulsar intensity function from measured photon time tags is called *epoch folding*. It is performed as follows: all the time tags during the observation time are collected. Then they are folded back into a single time interval equal to one pulse period, meaning that their modulus with one time period is calculated. Afterward, the period duration is divided into some equal-length bins and the number of photons in each bin is counted (see Fig. 3.5). Finally, the computed photon counts are normalized and the empirical pulsar profile is derived. Epoch folding is mathematically formulated in the rest of this section.

Assume that the observation time consists of N_P pulsar periods. In other words, $T_{\text{obs}} \approx N_p P$, where P is the pulsar period. Let $c(t_i)$ be the number of detected photons in the ith bin whose center is t_i. Also, assume that the bin size is T_b seconds and the number of time bins in one period is N_b, i.e., $T_b = P/N_\text{b}$. Using (3.8), for

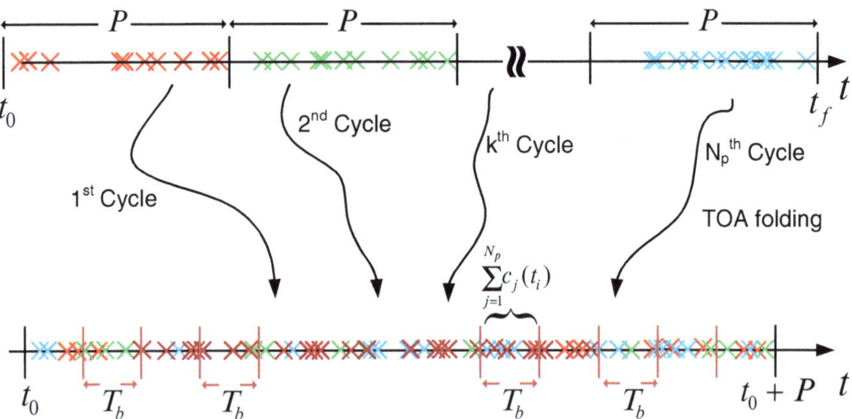

Fig. 3.5 Epoch folding: All [color-coded] photon TOAs are folded back into the first cycle $[t_0, t_0 + P]$, and it is divided to some equal-length bins. Then, the number of photons in each bin is counted and normalized

infinitesimal T_b values, $c(t_i)$ is a Poisson random variable with the following mean and variance,

$$\begin{aligned} E[c(t_i)] &= \text{var}[c(t_i)] \\ &= \lambda(t_i)T_b. \end{aligned} \tag{3.38}$$

The bin size, T_b, must be chosen small enough such that (3.38) is not violated. Now, consider the random variable $\breve{\lambda}(t_i)$, named empirical rate function, which represents the normalized number of photons resulted from folding back N_p epochs of time tags into the ith bin,

$$\breve{\lambda}(t_i) = \frac{1}{N_p T_b} \sum_{j=1}^{N_p} c_j(t_i). \tag{3.39}$$

Then, the relation between $\breve{\lambda}(t_i)$ and $\lambda(t_i)$ is given in the following theorem [43].

Theorem 3.2. *Let $\lambda(t_i)$, $0 \leq t_i \leq P$, be the true rate function, and $\breve{\lambda}(t_i)$ be the empirical rate function, obtained from an epoch folding procedure. Then,*

$$\breve{\lambda}(t_i) = \lambda(t_i) + \breve{n}(t_i) \quad \text{when} \quad T_{\text{obs}} \to \infty, \tag{3.40}$$

where $\breve{n}(t_i)$ is referred to as the epoch folding noise. *The noise, $\breve{n}(t_i)$, is uncorrelated and for long observation times, its mean and variance are given by*

$$E[\breve{n}(t_i)] = 0 \tag{3.41}$$

$$var[\breve{n}(t_i)] = \frac{N_\text{b}}{T_{\text{obs}}} \lambda(t_i). \tag{3.42}$$

Proof. Consider the average random variable, $\breve{C}(t_i)$, which obtained from folding back N_p epochs of time tags into ith bin

$$\breve{C}(t_i) = \frac{1}{N_p} \sum_{j=1}^{N_p} c_j(t_i), \tag{3.43}$$

where j is the epoch's index. Then, from (3.38), as $N_p \to \infty$ (or equivalently $T_{\text{obs}} \to \infty$),

$$E[\breve{C}(t_i)] = \lambda(t_i)T_b \tag{3.44}$$

$$\text{var}\left[\sqrt{N_p}\,\breve{C}(t_i)\right] = \lambda(t_i)T_b. \tag{3.45}$$

Therefore, using (3.39)

$$E[\breve{\lambda}(t_i)] = \lambda(t_i) \tag{3.46}$$

Fig. 3.6 Two photon TOAs

$$\begin{aligned}\mathrm{var}[\breve{\lambda}(t_i)] &= \frac{\lambda(t_i)}{N_p T_b}\\ &= \frac{N_{\mathrm{b}}}{T_{\mathrm{obs}}}\lambda(t_i).\end{aligned} \tag{3.47}$$

As a result of (3.46) and (3.47), the empirical rate function can be modeled by (3.40), (3.41), and (3.42).

To prove that $\breve{n}(t_i)$ is uncorrelated, the first and second moments of N_t are calculated using (3.8),

$$E[N_t] = \int_0^t \lambda(\tau)\mathrm{d}\tau \tag{3.48}$$

$$E[N_t^2] = \left(\int_0^t \lambda(\tau)\mathrm{d}\tau\right)^2 + \int_0^t \lambda(\tau)\mathrm{d}\tau. \tag{3.49}$$

Consider two time instants t_i and t_k, where $t_k > t_i$ (see Fig. 3.6). Because non-overlapping increments of N_t are independent,

$$\begin{aligned}E[N_{t_i}N_{t_k}] &= E[N_{t_i}(N_{t_k} - N_{t_i} + N_{t_i})]\\ &= E[N_{t_i}(N_{t_k} - N_{t_i})] + E[N_{t_i}^2]\\ &= E[N_{t_i}]E[N_{t_k} - N_{t_i}] + E[N_{t_i}^2].\end{aligned} \tag{3.50}$$

Then, using (3.48) and (3.49)

$$E[N_{t_i}N_{t_k}] = \int_0^{t_i} \lambda(\tau)\mathrm{d}\tau \left(\int_{t_i}^{t_k} \lambda(\tau)\mathrm{d}\tau + \int_0^{t_i} \lambda(\tau)\mathrm{d}\tau + 1\right). \tag{3.51}$$

For infinitesimal values of T_b, the number of detected photons in ith bin, i.e., $c(t_i)$, can be approximated by

$$c(t_i) \simeq N_{t_i+T_b/2} - N_{t_i-T_b/2}. \tag{3.52}$$

Hence,

$$\begin{aligned}E[c(t_i)c(t_k)] &= E[(N_{t_i+T_b/2} - N_{t_i-T_b/2})(N_{t_k+T_b/2} - N_{t_k-T_b/2})]\\ &= E[N_{t_i+T_b/2}N_{t_k+T_b/2}] - E[N_{t_i+T_b/2}N_{t_k-T_b/2}]\\ &\quad - E[N_{t_i-T_b/2}N_{t_k+T_b/2}] + E[N_{t_i-T_b/2}N_{t_k-T_b/2}].\end{aligned} \tag{3.53}$$

Calculating each term in (3.53), using (3.51), it results in

$$\begin{aligned}
E[c(t_i)c(t_k)] &= \int_{t_k-T_b/2}^{t_k+T_b/2} \lambda(\tau)\mathrm{d}\tau \left(\int_0^{t_i+T_b/2} \lambda(\tau)\mathrm{d}\tau - \int_0^{t_i-T_b/2} \lambda(\tau)\mathrm{d}\tau \right) \\
&= \int_{t_k-T_b/2}^{t_k+T_b/2} \lambda(\tau)\mathrm{d}\tau \int_{t_i-T_b/2}^{t_i+T_b/2} \lambda(\tau)\mathrm{d}\tau \qquad (3.54)
\end{aligned}$$

which can be approximated by

$$E[c(t_i)c(t_k)] \approx T_b^2 \lambda(t_i)\lambda(t_k) \qquad (3.55)$$

for infinitesimal values of T_b.

As a result, the auto-correlation function of $\breve{\lambda}(t_i)$ can be calculated from (3.39)

$$\begin{aligned}
E[\breve{\lambda}(t_i)\breve{\lambda}(t_k)] &= \left(\frac{1}{N_p T_b}\right)^2 E\left[\sum_{j=1}^{N_p} c_j(t_i) \sum_{j=1}^{N_p} c_j(t_k)\right] \\
&= \left(\frac{1}{N_p T_b}\right)^2 \sum_{j=1}^{N_p} \sum_{r=1}^{N_p} E[c_j(t_i)c_r(t_k)] \\
&= \lambda(t_i)\lambda(t_k). \qquad (3.56)
\end{aligned}$$

On the other hand, noting that $\breve{n}(t)$ is zero mean, (3.40) leads to

$$E[\breve{\lambda}(t_i)\breve{\lambda}(t_k)] = \lambda(t_i)\lambda(t_k) + E[\breve{n}(t_i)\breve{n}(t_k)]. \qquad (3.57)$$

Then, from (3.56) and (3.57), the auto correlation function of $\breve{n}(t)$ is given by

$$E[\breve{n}(t_i)\breve{n}(t_k)] = 0. \qquad (3.58)$$

■

To perform epoch folding, the observed pulsar frequency, $f_o = 1/P$, must be constant and known during the observation time. In other words, the detector velocity, v, must be a known constant. Since in practice the velocity is not perfectly known, the effect of velocity errors on epoch folding is studied in the next section.

3.7 Epoch Folding in Presence of Velocity Errors

As mentioned, it is necessary to investigate how much the epoch folding procedure is tolerant to possible errors in the spacecraft velocity data. Let the detector velocity error be Δv. From (3.26), this error causes the observed frequency to change as

$$\Delta f_o = f_s \frac{\Delta v}{c}. \qquad (3.59)$$

Since $P = 1/f_o$, Δf_o makes P change by

$$\Delta P = -\frac{\Delta f_o}{f_o^2} \tag{3.60}$$

which can be simplified using (3.59), as

$$\Delta P = -\frac{f_s}{c f_o^2}\Delta v. \tag{3.61}$$

Hence, the observed pulsar period will be

$$\check{P} = P + \Delta P. \tag{3.62}$$

This pulsar period, which is ΔP seconds off from the correct one, will be used to fold back the photon TOAs into one cycle (see Fig. 3.7). Assuming the number of bins is still N_b, the new bin size is given by

$$\check{T}_b = \frac{\check{P}}{N_b}. \tag{3.63}$$

Furthermore, the number of epochs to be folded back is

$$\check{N}_p = \frac{T_{obs}}{\check{P}}. \tag{3.64}$$

The error ΔP, can be translated to the phase domain

$$\begin{aligned} \Delta\phi &= \frac{\Delta P}{P} \\ &= f_o \Delta P \end{aligned} \tag{3.65}$$

and using (3.61), it can be simplified by

$$\begin{aligned} \Delta\phi &= -\frac{f_s}{c f_o}\Delta v \\ &\approx -\frac{1}{c}\Delta v. \end{aligned} \tag{3.66}$$

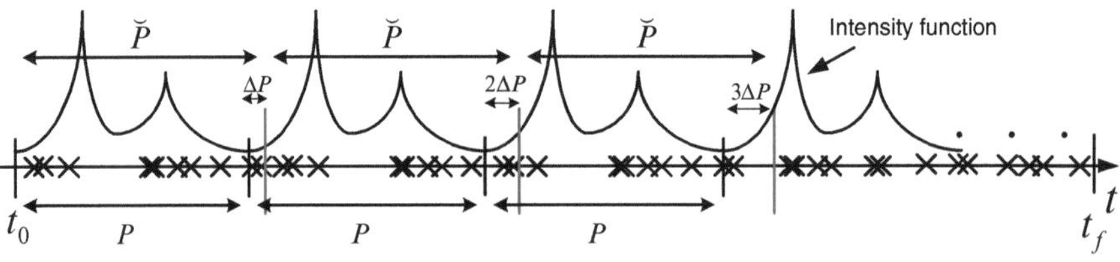

Fig. 3.7 Epoch folding in the presence of the spacecraft velocity error

From (3.38), this bias error in phase changes the statistical properties of $c_j(t_i)$ as follows.

$$\begin{aligned} E[c_j(t_i)] &= \text{var}[c_j(t_i)] \\ &= \lambda\left(t_i;\phi_0 + (j-1)\Delta\phi\right)\check{T}_b, \quad j = 1,2,\ldots,\check{N}_p \end{aligned} \tag{3.67}$$

In other words, the mean of folded back photon counts from the jth cycle is shifted by $(j-1)\Delta\phi$ cycle. Thus, the empirical rate function equals the following [44].

$$\begin{aligned} \check{\lambda}(t_i) &= \frac{1}{\check{N}_p\check{T}_b}\sum_{j=1}^{\check{N}_p} c_j(t_i) \\ &= \frac{1}{\check{N}_p}\sum_{j=1}^{\check{N}_p} \lambda\left(t_i;\phi_0 + (j-1)\Delta\phi\right) + \check{n}(t_i), \end{aligned} \tag{3.68}$$

where

$$E[\check{n}(t_i)] = 0 \tag{3.69}$$

and

$$\text{var}[\check{n}(t_i)] = \frac{1}{\check{N}_p^2\check{T}_b}\sum_{j=1}^{\check{N}_p} \lambda\left(t_i;\phi_0 + (j-1)\Delta\phi\right). \tag{3.70}$$

From (3.70) it is clear that if $(j-1)\Delta\phi$ is always smaller than the bin size, (3.70) boils down to the result given in (3.40). That is, if

$$|\Delta\phi|(\check{N}_p - 1) \approx |\Delta\phi|\check{N}_p < \frac{1}{N_b} \tag{3.71}$$

or, using (3.66)

$$\Delta v < \frac{c}{N_b\check{N}_p} \tag{3.72}$$

then, the empirical function satisfactorily represents the true rate function. The upper bound in (3.72) shows how much error on the spacecraft velocity data can be tolerated by the epoch folding procedure. If the velocity error exceeds the bound, as (3.70) shows, the empirical rate function will be a deteriorated version of the true function. The deterioration shows as a shift in the initial phase, and a change in the magnitude of the rate function.

An interesting point (3.72) shows is that the upper bound is inversely proportional to the number of bins N_b. But one should be careful in choosing N_b too small, because (3.38) must not be violated.

3.8 Generating Photon TOAs

To assess the analytical results, the X-ray photon TOAs corresponding to a particular known pulsar rate function $\lambda(t)$ must be simulated. To realize corresponding TOAs, an algorithm based on inversion of the integrated rate function is used [43, 45].

Let U be a uniform random variable in the interval $(0,1)$. Define

$$g(u) = F_y^{-1}(u), \tag{3.73}$$

where $F_y^{-1}(\cdot)$ is the inverse function of $F_y(\cdot)$. Then, $Y = g(U)$ is a random variable with the distribution $F_y(y)$ [46]. That is,

$$\text{If } y = F_y^{-1}(U) \text{ then } P\{Y \leq y\} = F_y(y). \tag{3.74}$$

Now let $F_{z|t_n=t}(z|t_n = t)$ be the probability distribution of $Z = t_{n+1} - t_n$ given $t_n = t$, where t_i is the TOA of the ith photon. Then,

$$P(Z > z|t_n = t) = 1 - F_{z|t_n=t}(z|t_n = t). \tag{3.75}$$

Also, using (3.9)

$$\begin{aligned} P(Z > z|t_n = t) &= P(N_{t+z} - N_t = 0) \\ &= \exp\left(-\int_t^{t+z} \lambda(t)\mathrm{d}t\right) \\ &= \exp\Big(-\big(\Lambda(t+z) - \Lambda(t)\big)\Big), \end{aligned} \tag{3.76}$$

where $\Lambda(t)$ is defined in (3.8). Therefore, from (3.75) and (3.76)

$$F_{z|t_n=t}(z|t_n = t) = 1 - \exp\{-\big(\Lambda(t+z) - \Lambda(t)\big)\} \tag{3.77}$$

and

$$F_{z|t_n=t}^{-1}(z|t_n = t) = -t + \Lambda^{-1}\big(\Lambda(t) - \ln(1-z),\big) \tag{3.78}$$

where $F_{z|t_n=t}^{-1}(\cdot)$ is the inverse function of $F_{z|t_n=t}(\cdot)$. Hence, given $t_n = t$, Z can be generated as

$$Z = -t + \Lambda^{-1}\big(\Lambda(t) - \ln(1-U)\big). \tag{3.79}$$

As $1-U$ is also a uniform random variable in the interval $(0,1)$, without loss of generality, (3.79) can be restated as

$$Z = -t + \Lambda^{-1}\big(\Lambda(t) - \ln U\big). \tag{3.80}$$

Also, note that if X is an exponential random variable, described by the following pdf,

$$p(x) = \begin{cases} \lambda_e \mathrm{e}^{-\lambda_e x} & x \geq 0 \\ 0 & \text{otherwise} \end{cases} \tag{3.81}$$

then, its probability distribution is given by

$$F_x(x) = 1 - \mathrm{e}^{-\lambda_e x} \tag{3.82}$$

and

$$F_x^{-1}(x) = -\frac{1}{\lambda_e} \ln(1-x). \tag{3.83}$$

Therefore, from (3.74)

$$X = -\frac{1}{\lambda_e} \ln(1-U) \tag{3.84}$$

is an exponential random variable with parameter λ_e. As $1-U$ is also a uniform random variable in the interval $(0,1)$, it can be concluded that

$$X = -\frac{1}{\lambda_e} \ln U \tag{3.85}$$

is an exponential random variable with parameter λ_e. Thus, the term $-\ln U$ in (3.80) is an exponential random variable, named E, with parameter $\lambda_e = 1$. As a result, using (3.80), given $t = t_n$, t_{n+1} can be generated as

$$\begin{aligned} t_{n+1} &= t_n + Z \\ &= \Lambda^{-1}\big(\Lambda(t_n) + E\big). \end{aligned} \tag{3.86}$$

The preceding discussion is concluded into Algorithm 3.1.

Algorithm 3.1 *TOA Simulation:*

- $L \leftarrow 0$ (L is an auxiliary variable)
- $k \leftarrow 0$ (k is a counter)
- WHILE $t_k \leq t_f$
  - Generate an exponential random variable, E, with parameter $\lambda_e = 1$.
  - $k \leftarrow k+1$
  - $L \leftarrow \Lambda^{-1}\big(\Lambda(L) + E\big)$
  - $t_k \leftarrow L$
- END

3.9 Numerical Examples

The Crab pulsar (PSR B0531+21) is selected to verify the analytical results. The pulsar period is 33.5 ms, and its profile, $h(\phi)$, is shown in Fig. 3.8. The constant arrival rates are chosen to be $\lambda_b = 5$ and $\lambda_s = 15$ ph/s. The detector's velocity is assumed to be $v = 3$ km/s; and the initial phase observed at the first detector is $\phi_0 = 0.16$ cycle.

A Monte Carlo simulation was performed in which 500 independent realization of photon TOAs were processed. Photon TOAs were generated as realizations of a non-homogeneous Poisson process, as explained in Algorithm 3.1. The accumulated rate function used to generate the TOAs, Λ, is plotted in Fig. 3.8.

An observation time of $T_{obs} = 100$ s is selected to verify the epoch folding results. The photon time tags are folded back into one single pulsar cycle which is divided into $N_b = 1{,}024$ bins, and the rate functions on each detector are derived using (3.39). The empirical rate function along with the true one is plotted in Fig. 3.9. As the plots show, the empirical rate function converges to the true function.

In Fig. 3.9, the variance of $\breve{n}(t)$, obtained via simulation, is also plotted and compared to the analytical variance for each detector. The plots show an excellent

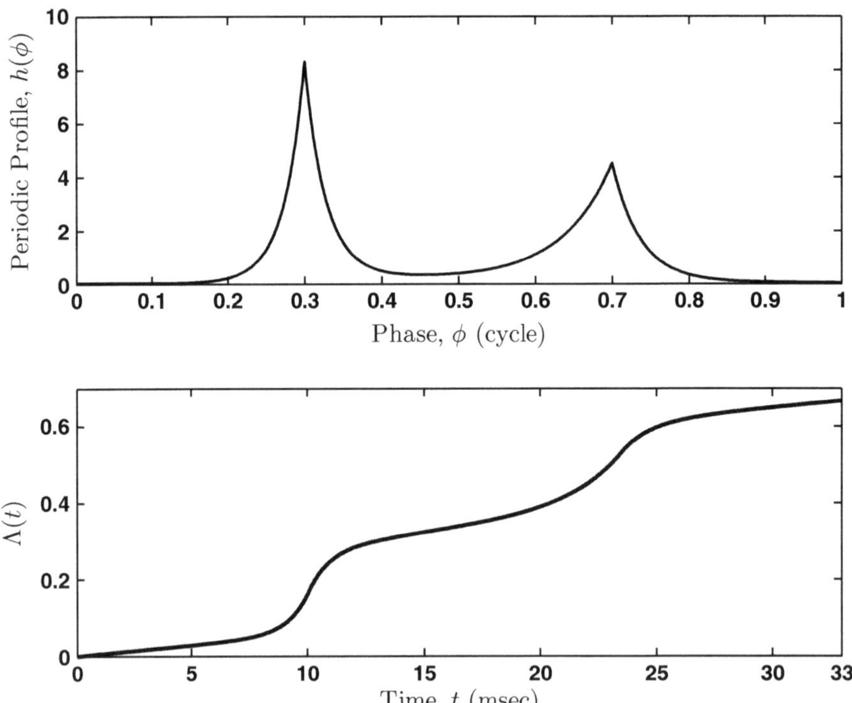

Fig. 3.8 Crab pulsar profile, and $\Lambda(t)$ for $\lambda_b = 5$ and $\lambda_s = 15$ ph/s

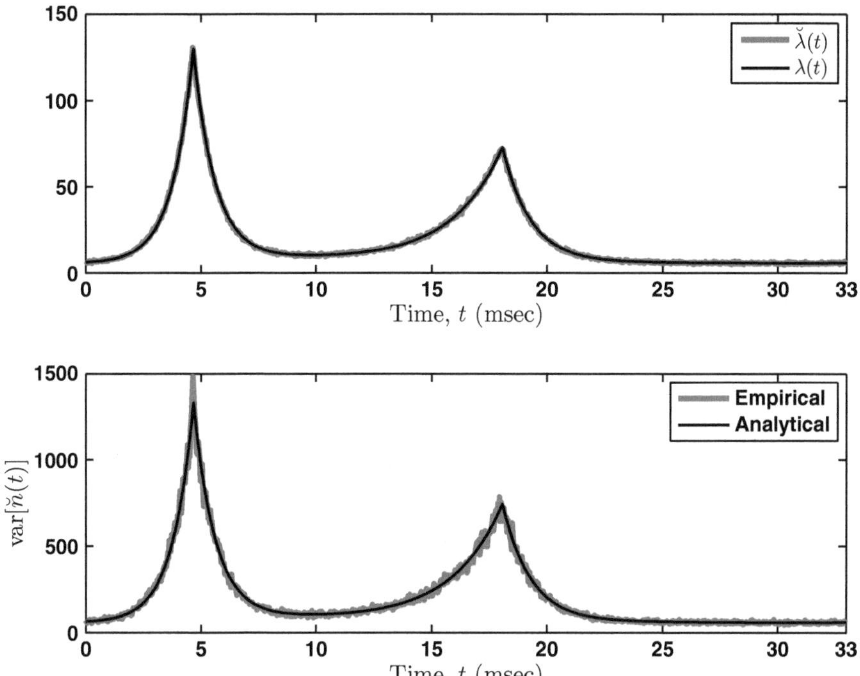

Fig. 3.9 Pulsar rate function, and the variance of $\breve{n}(t)$ for $T_{\text{obs}} = 100\,\text{s}$

agreement between the simulation and the analytical value, given in (3.42). The difference between these two plots gets larger at the time instants where there is a sharp change in the rate function. This phenomenon is due to the limited number of time bins and finite observation time which violate (3.38). In other words, if the observation time is infinitely long and the time bins are infinitely small then these jumps in empirical graphs will vanish.

The auto-correlation of $\breve{\lambda}(t)$, $E[\breve{\lambda}(t)\breve{\lambda}(t+D_0)]$, is calculated numerically for the time delay $D_0 = P/4$ second, where P is the pulsar observed period. The plot is shown in Fig. 3.10, and it verifies that $E[\breve{\lambda}(t)\breve{\lambda}(t+D_0)] = \lambda(t)\lambda(t+D_0)$. This means that the epoch folding noise, $\breve{n}(t)$, is uncorrelated.

The effect of spacecraft imprecise absolute velocity data on epoch folding is studied as well. The velocity error is intentionally chosen to be big: $\Delta v = 1{,}000$ m/s. For the selected parameters, the upper bound (3.72) roughly equals 10 m/s. Since the velocity error is bigger than the bound, the empirical rate function and the noise variance are expected to be deteriorated as opposed to the case where the velocity is perfectly known. Figure 3.11 verifies this fact. Comparing the empirical rate function to the true one shows that it is shifted, and its peak is shortened and flattened. The figure also verifies that the variance of the epoch folding noise is shifted because of the velocity error.

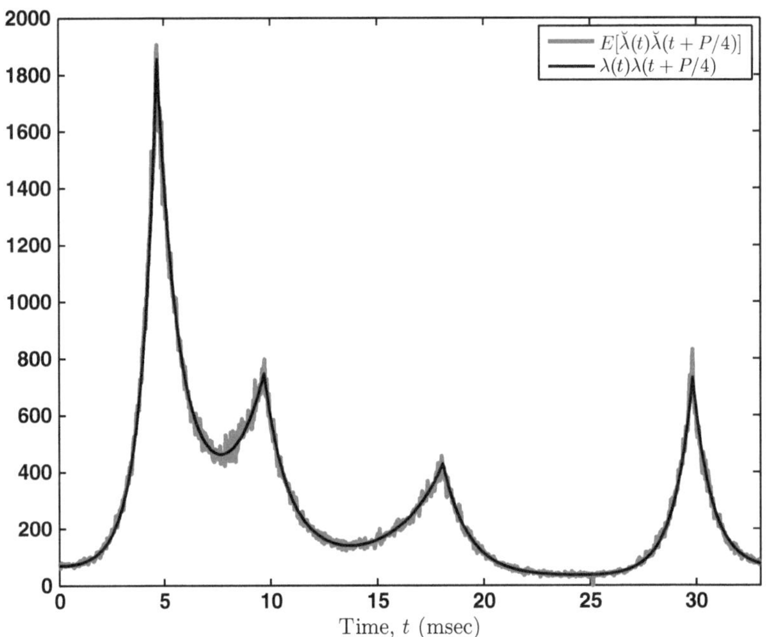

Fig. 3.10 Autocorrelation of $\breve{\lambda}(t)$ for $D_0 = P/4$

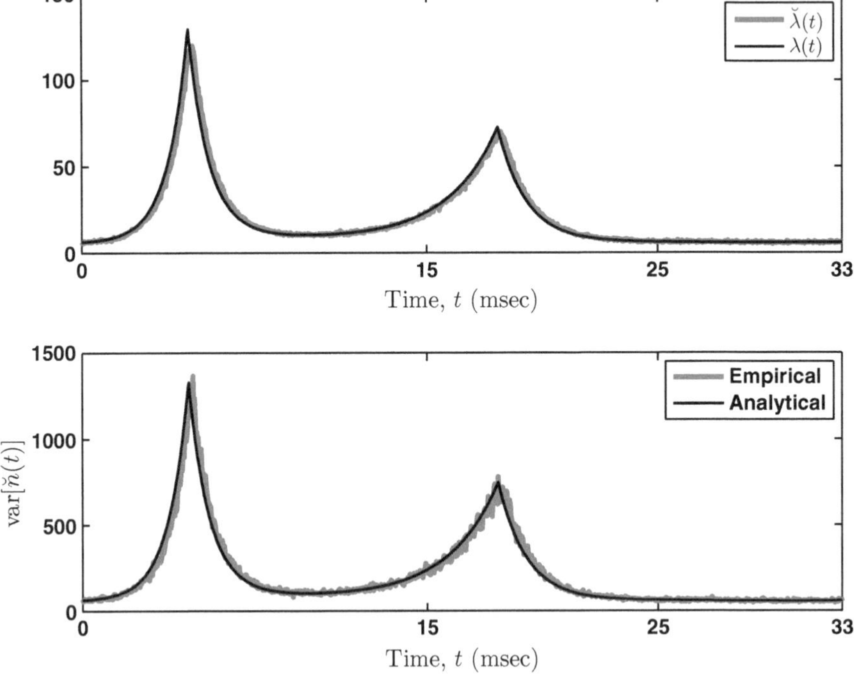

Fig. 3.11 $\breve{\lambda}(t)$ and the variance of $\breve{n}(t)$ when $\Delta\nu = 1{,}000$ m/s

3.10 Summary

In this chapter, we offer an overview of the proposed navigation system structure. We explain how by estimation of the time delay between the X-ray pulsar signals, the spacecraft relative or absolute position can be estimated. More detailed mathematical formulation of the estimation problem, and our proposed solutions to the problem will be presented in the following chapters. We also suggest using IMUs to provide the spacecraft acceleration data. We explain how based on time tagging the photon time of arrivals, X-ray pulsar detectors work. Using the presented detection mechanism, we develop the mathematical models needed to characterize the pulsar signals. We formulate and analyze the epoch folding procedure and show how it results in recovering the pulsar intensity function. We investigate how errors on the spacecraft velocity data can affect the epoch folding, and show that they result in recovering a deteriorated rate function. Finally, we examine the theoretical results via numerical simulations.

Chapter 4
Pulse Delay Estimation

4.1 Introduction

As explained in Chap. 3, the navigation system measurement is obtained through estimation of the time delay between the received signals. The delay estimation problem plays the most important role in the navigation system. In this chapter, we study this problem in more detail.

Section 4.2 offers a mathematical formulation of the delay estimation problem. In Sects. 4.3 and 4.4, we present a lower bound on the variance of the estimation error, called the Cramér–Rao lower bound (CRLB). Some numerical examples on calculation of the CRLB is also given in Sect. 4.5.

4.2 Pulse Delay Estimation

Let $\{t_i^{(1)}\}_{i=1}^{M_1}$ be the measured photon TOAs at the first detector. This TOA set corresponds to the following rate function.

$$\lambda_1(t;\phi_1) = \lambda_b + \lambda_s h\big(\phi_1 + (t - t_0)f_1\big) \tag{4.1}$$

At the second spacecraft, a different TOA set is measured, $\{t_i^{(2)}\}_{i=1}^{M_2}$, which corresponds to the delayed rate function observed by the first detector

$$\lambda_2(t;\phi_1) = \lambda_b + \lambda_s h\big(\phi_1 + (t - t_0 - t_e - t_x)f_2\big), \tag{4.2}$$

where t_x is the true time delay between the detected rate functions, and t_e is the differential time between clocks [47]. Let $t_d = t_e + t_x$, and define ϕ_2 as

$$\phi_2 \triangleq \phi_1 - f_2 t_d. \tag{4.3}$$

A.A. Emadzadeh and J.L. Speyer, *Navigation in Space by X-ray Pulsars*,
DOI 10.1007/978-1-4419-8017-5_4, © Springer Science+Business Media, LLC 2011

Then, the rate function $\lambda_2(\cdot)$ in (4.2) can be simplified as

$$\lambda_2(t;\phi_2) = \lambda_b + \lambda_s h\big(\phi_2 + (t - t_0) f_2\big). \tag{4.4}$$

Note that f_1 and f_2 are observed pulsar frequencies. To construct the measurements for the Kalman filter, t_d must be estimated.

To estimate the pulse delay, t_d, ϕ_1 and ϕ_2 will be estimated on each detector. Then, using (4.3), t_d can be found. We have proposed two different methods for estimation of the pulsar intensity function's initial phase, and analyzed their characteristics.

1. Recovering the pulsar rate functions at each detector through the epoch folding procedure represents the first strategy. The empirical rate function will be used for estimation of the initial phase values. Two different pulse delay estimators based on epoch folding are introduced, and their performance against the CRLB is analyzed. One uses the cross-correlation function between the empirical and true photon intensity functions [44]. The other employs a least-squares criterion [43, 44, 48]. Details are given in Chap. 5.
2. The second strategy is based on the direct use of the measured photon TOAs. To estimate the unknown parameters (initial phase values and Doppler frequencies), a maximum likelihood (ML) estimation problem is formulated using the pdf of the photon time tags [43, 49]. Different aspects of this approach are explained in Chap. 6.

4.3 CRLB

Being able to place a lower bound on the variance of any unbiased estimator is extremely useful in practice. It allows one to assert if the estimator is the minimum variance estimator and provides a benchmark against which the performance of any unbiased estimator can be compared. An estimator that is unbiased and attains the CRLB is said to be *efficient*, meaning that it efficiently uses the data. The CRLB is stated in the following theorem [50].

Theorem 4.1. *Let $p(\boldsymbol{x};\theta)$ be the probability density function (or probability mass function) of the observed random vector, $\boldsymbol{x}$, which is parameterized by the unknown parameter θ. A semicolon is used to denote the dependence. Assume that $p(\boldsymbol{x};\theta)$ satisfies the following "regularity" conditions*

$$E\left[\frac{\partial}{\partial\theta}\ln p(\boldsymbol{x};\theta)\right] = 0, \quad \textit{for all } \theta, \tag{4.5}$$

where θ is the unknown parameter vector, and the expectation is taken with respect to $p(\boldsymbol{x};\theta)$. Then, the covariance matrix of any unbiased estimator $\hat{\theta}$ satisfies

$$cov(\hat{\theta}) \geq I^{-1}(\theta), \tag{4.6}$$

where the inequality "$\geq$" means $\left(cov(\hat{\theta}) - I^{-1}(\theta)\right)$ is positive semidefinite and $I(\theta)$ is the Fisher matrix given by

$$[I(\theta)]_{ij} = -E\left[\frac{\partial^2}{\partial\theta_i\partial\theta_j}\ln p(\boldsymbol{x};\theta)\right], \tag{4.7}$$

where the derivatives are evaluated at the true value of θ and the expectation is taken with respect to $p(\boldsymbol{x};\theta)$. The matrix $I^{-1}(\theta)$ is called the Cramér–Rao lower bound.

Now consider the following pulsar rate function, which corresponds to the measured TOA set on the detector.

$$\lambda(t;\phi_0,f_o) = \lambda_b + \lambda_s h\big(\phi_0 + (t-t_0)f_o\big). \tag{4.8}$$

Assuming that ϕ_0 and f_o are not known, the aim is to find the CRLB for estimation of these parameters. The result is presented in the following theorem [51,52].

Theorem 4.2. *Let*

$$\theta \triangleq \left(\phi_0 \; f_o\right)^{\mathrm{T}} \tag{4.9}$$

be the unknown parameters vector. The Fisher matrix and its inverse (CRLB) for estimation of θ in (4.8) are given by

$$I(\theta) = \frac{I_p}{6}\begin{pmatrix} 6T_{\text{obs}} & 3T_{\text{obs}}^2 \\ 3T_{\text{obs}}^2 & 2T_{\text{obs}}^3 \end{pmatrix}, \tag{4.10}$$

where

$$I_p \triangleq \int_0^1 \frac{[\lambda_s h'(\phi)]^2}{\lambda_b + \lambda_s h(\phi)}\,\mathrm{d}\phi. \tag{4.11}$$

Therefore,

$$\begin{aligned} \text{CRLB}(\theta) &= I^{-1}(\theta) \\ &= \frac{2}{I_p}\begin{pmatrix} 2/T_{\text{obs}} & -3/T_{\text{obs}}^2 \\ -3/T_{\text{obs}}^2 & 6/T_{\text{obs}}^3 \end{pmatrix}. \end{aligned} \tag{4.12}$$

Proof. Let θ_1 and θ_2 be the elements of the vector θ (this notation is used for simplicity). To derive the CRLB, the photon TOA measurements are transformed into another set of data containing the number of detected photons. This new data set is formed by dividing the observation time into N number of bins whose length is Δt and by placing the number of counted photons in each bin into the vector

$$\mathbf{x} = \left(x_1 \; x_2 \; \cdots \; x_N\right)^{\mathrm{T}} \tag{4.13}$$

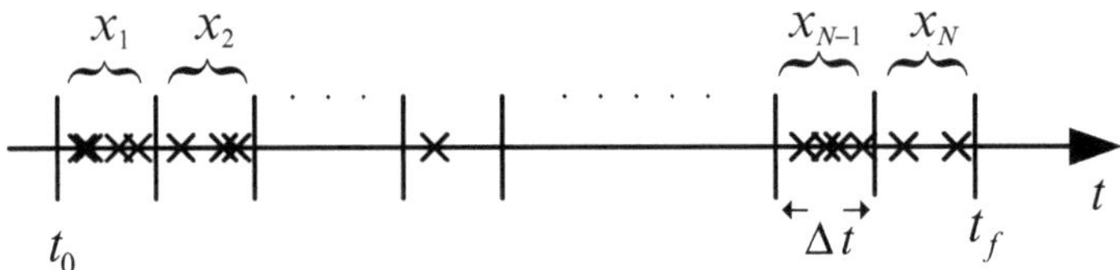

Fig. 4.1 Binning the observation time

as shown in Fig. 4.1. For sufficiently small Δt, the rate of arrival in nth bin can be assumed to be constant:

$$\lambda_n(\theta) \approx \frac{1}{\Delta t} \int_{t_0+(n-1)\Delta t}^{t_0+n\Delta t} \lambda(t;\theta)\mathrm{d}t. \tag{4.14}$$

The bin count x_n is a Poisson random variable whose probability is presented by

$$p(x_n = k;\theta) = \frac{\left(\lambda_n(\theta)\Delta t\right)^k}{k!} \exp\left(-\lambda_n(\theta)\Delta t\right), \quad k = 0,1,2,\ldots \tag{4.15}$$

and its mean and variance are

$$\begin{aligned} E\left[x_n;\theta\right] &= \mathrm{var}(x_n;\theta) \\ &= \lambda_n(\theta)\Delta t. \end{aligned} \tag{4.16}$$

As x_n random variables are independent, their joint probability mass function (pmf) equals

$$\begin{aligned} p(\mathbf{x};\theta) &= \prod_{n=1}^{N} p(x_n;\theta) \\ &= \prod_{n=1}^{N} \frac{\left(\lambda_n(\theta)\Delta t\right)^{x_n}}{x_n!} \exp\left(-\lambda_n(\theta)\Delta t\right). \end{aligned} \tag{4.17}$$

To derive the CRLB using (4.17), first, it must be checked if the regularity condition (4.5) holds. Therefore, the natural logarithm of (4.17) must be calculated

$$\ln p(\mathbf{x};\theta) = \sum_{n=0}^{N-1} \left(x_n \ln[\lambda_n(\theta)\Delta t] - \lambda_n(\theta)\Delta t - \ln(x_n!)\right). \tag{4.18}$$

The derivative of (4.18) with respect to θ equals

$$\frac{\partial}{\partial\theta} \ln p(\mathbf{x};\theta) = \sum_{n=0}^{N-1} \left(x_n \cdot \frac{1}{\lambda_n(\theta)\Delta t} \cdot \frac{\partial}{\partial\theta}\lambda_n(\theta)\Delta t - \frac{\partial}{\partial\theta}\lambda_n(\theta)\Delta t\right) \tag{4.19}$$

and its expectation is given by

$$E\left[\frac{\partial}{\partial\theta}\ln p(\mathbf{x};\theta)\right] = \sum_{n=0}^{N-1}\left(E[x_n]\cdot\frac{1}{\lambda_n(\theta)\Delta t}\cdot\frac{\partial}{\partial\theta}\lambda_n(\theta)\Delta t - \frac{\partial}{\partial\theta}\lambda_n(\theta)\Delta t\right)$$
$$= \sum_{n=0}^{N-1}\left(\lambda_n(\theta)\Delta t\cdot\frac{1}{\lambda_n(\theta)\Delta t}\cdot\frac{\partial}{\partial\theta}\lambda_n(\theta)\Delta t - \frac{\partial}{\partial\theta}\lambda_n(\theta)\Delta t\right)$$
$$= 0. \tag{4.20}$$

Hence, the regularity conditions (4.5) hold.

To obtain the CRLB, the Fisher matrix $I(\theta)$ must be found. From (4.7) and (4.16), the matrix elements, I_{ij}, are given by

$$I_{ij} = -E\left[\frac{\partial^2}{\partial\theta_i\partial\theta_j}\ln p(\mathbf{x};\theta)\right]$$
$$= \sum_{n=0}^{N-1}\left(-E[x_n]\cdot\frac{\partial^2}{\partial\theta_i\partial\theta_j}\ln[\lambda_n(\theta)\Delta t] + \frac{\partial^2}{\partial\theta_i\partial\theta_j}\lambda_n(\theta)\Delta t\right)$$
$$= \sum_{n=0}^{N-1}\left(-(\lambda_n(\theta)\Delta t)\left\{\frac{1}{\lambda_n(\theta)\Delta t}\cdot\frac{\partial^2}{\partial\theta_i\partial\theta_j}\lambda_n(\theta)\Delta t\right.\right.$$
$$\left.\left. - \left[\frac{1}{\lambda_n(\theta)\Delta t}\right]^2\cdot\frac{\partial}{\partial\theta_i}\lambda_n(\theta)\Delta t\cdot\frac{\partial}{\partial\theta_j}\lambda_n(\theta)\Delta t\right\} + \frac{\partial^2}{\partial\theta_i\partial\theta_j}\lambda_n(\theta)\Delta t\right)$$
$$= \sum_{n=0}^{N-1}\frac{1}{\lambda_n(\theta)}\cdot\frac{\partial}{\partial\theta_i}\lambda_n(\theta)\cdot\frac{\partial}{\partial\theta_j}\lambda_n(\theta)\cdot\Delta t. \tag{4.21}$$

By letting $\Delta t \to 0$, the summation in (4.21) can be replaced by an integral

$$I_{ij} = \int_{t_0}^{t_f}\frac{\frac{\partial}{\partial\theta_i}\lambda(t;\theta)\cdot\frac{\partial}{\partial\theta_j}\lambda(t;\theta)}{\lambda(t;\theta)}\mathrm{d}t, \quad i,j = 1,2. \tag{4.22}$$

Using (4.8), the partial derivatives in (4.22) are

$$\frac{\partial}{\partial\theta_1}\lambda(t;\theta) = \frac{\partial}{\partial\phi_0}\lambda(t;\theta) = \lambda_s\cdot h'\big(\phi_0 + (t-t_0)f_o\big) \tag{4.23}$$

$$\frac{\partial}{\partial\theta_2}\lambda(t;\theta) = \frac{\partial}{\partial f_o}\lambda(t;\theta) = \lambda_s\cdot(t-t_0)\cdot h'\big(\phi_0 + (t-t_0)f_o\big). \tag{4.24}$$

Substituting (4.23) into (4.22), I_{11} may be found,

$$
\begin{aligned}
I_{11} &= \int_{t_0}^{t_f} \frac{\left[\frac{\partial}{\partial \phi_0}\lambda(t;\theta)\right]^2}{\lambda(t;\theta)} \mathrm{d}t \\
&= \int_{t_0}^{t_0+T_{\text{obs}}} \frac{\left[\lambda_s h'(\phi_0 + (t-t_0)f_o)\right]^2}{\lambda_b + \lambda_s h(\phi_0 + (t-t_0)f_o)} \mathrm{d}t \\
&= \int_0^{T_{\text{obs}}} \frac{\left[\lambda_s h'(\phi_0 + f_o t)\right]^2}{\lambda_b + \lambda_s h(\phi_0 + f_o t)} \mathrm{d}t. \qquad (4.25)
\end{aligned}
$$

The observation time can decomposed as $T_{\text{obs}} = N_p \cdot P + T_p$, where P is the pulsar period, and $0 \le T_p < P$. Hence, the integral in (4.25) can be split as follows

$$
\begin{aligned}
I_{11} = &\sum_{n=1}^{N_p} \left(\int_{(n-1)P}^{nP} \frac{\left[\lambda_s h'(\phi_0 + f_o t)\right]^2}{\lambda_b + \lambda_s h(\phi_0 + f_o t)} \mathrm{d}t \right) \\
&+ \int_{N_p P}^{N_p P + T_p} \frac{\left[\lambda_s h'(\phi_0 + f_o t)\right]^2}{\lambda_b + \lambda_s h(\phi_0 + f_o t)} \mathrm{d}t. \qquad (4.26)
\end{aligned}
$$

If the observation time is long enough, i.e., $T_{\text{obs}} \gg 0$ then $T_{\text{obs}} \approx N_p \cdot P$ and the last integral in (4.26) almost equals to zero. Also, because $h(\cdot)$ is periodic, all of the remained integrals in (4.26) are equal. Therefore,

$$
\begin{aligned}
I_{11} &\approx \sum_{n=1}^{N_p} \int_{(n-1)P}^{nP} \frac{\left[\lambda_s h'(\phi_0 + f_o t)\right]^2}{\lambda_b + \lambda_s h(\phi_0 + f_o t)} \mathrm{d}t \\
&= N_p \int_0^P \frac{\left[\lambda_s h'(\phi_0 + f_o t)\right]^2}{\lambda_b + \lambda_s h(\phi_0 + f_o t)} \mathrm{d}t. \qquad (4.27)
\end{aligned}
$$

By changing the integration variable, and noting that $f_o = 1/P$ and $T_{\text{obs}} \approx N_p P$

$$
\begin{aligned}
I_{11} &= \frac{N_p}{f} \int_{\phi_0}^{\phi_0 + fP} \frac{\left[\lambda_s h'(\phi)\right]^2}{\lambda_b + \lambda_s h(\phi)} \mathrm{d}\phi \\
&\approx T_{\text{obs}} \int_{\phi_0}^{\phi_0 + 1} \frac{\left[\lambda_s h'(\phi)\right]^2}{\lambda_b + \lambda_s h(\phi)} \mathrm{d}\phi \\
&= T_{\text{obs}} \int_0^1 \frac{\left[\lambda_s h'(\phi)\right]^2}{\lambda_b + \lambda_s h(\phi)} \mathrm{d}\phi. \qquad (4.28)
\end{aligned}
$$

To find I_{22}, (4.24) is substituted into (4.7), the integration variable is changed, and the integral is split into summation of the integrals in each period

$$
\begin{aligned}
I_{22} &= \int_{t_0}^{t_0+T_{\text{obs}}} \frac{\left[\dfrac{\partial}{\partial f}\lambda(t;\theta)\right]^2}{\lambda(t;\theta)} \mathrm{d}t \\
&= \int_{t_0}^{t_0+T_{\text{obs}}} \frac{\left[\lambda_s(t-t_0)h'(\phi_0+(t-t_0)f_o)\right]^2}{\lambda_b+\lambda_s h(\phi_0+(t-t_0)f_o)} \mathrm{d}t \qquad (4.29) \\
&= \int_{0}^{T_{\text{obs}}} \frac{\left[\lambda_s\cdot t\cdot h'(\phi_0+f_o t)\right]^2}{\lambda_b+\lambda_s h(\phi_0+f_o t)} \mathrm{d}t \\
&= \sum_{n=1}^{N_p}\left(\int_{(n-1)P}^{nP} \frac{\left[\lambda_s\cdot t\cdot h'(\phi_0+f_o t)\right]^2}{\lambda_b+\lambda_s h(\phi_0+f_o t)} \mathrm{d}t\right) \\
&\quad + \int_{N_pP}^{N_pP+T_p} \frac{\left[\lambda_s\cdot t\cdot h'(\phi_0+f_o t)\right]^2}{\lambda_b+\lambda_s h(\phi_0+f_o t)} \mathrm{d}t. \qquad (4.30)
\end{aligned}
$$

As the observation time is long enough, the last integral in (4.30) is negligible. Also, notice that if $(n-1)P \le t \le nP$, then t in (4.30) can be replaced by

$$t = (n-1+\phi)P, \qquad (4.31)$$

where $0 \le \phi \le 1$. Hence,

$$
\begin{aligned}
I_{22} &\approx \sum_{n=1}^{N_p}\int_{(n-1)P}^{nP} \frac{\left[\lambda_s\cdot t\cdot h'(\phi_0+f_o t)\right]^2}{\lambda_b+\lambda_s h(\phi_0+f_o t)} \mathrm{d}t \\
&= \sum_{n=1}^{N_p}\int_{0}^{1} \frac{\left[\lambda_s\cdot(n-1+\phi)P\cdot h'(\phi_0+(n-1+\phi)f_oP)\right]^2}{\lambda_b+\lambda_s h(\phi_0+(n-1+\phi)f_oP)} P\mathrm{d}\phi \\
&= P^3\sum_{n=1}^{N_p}\int_{0}^{1} \frac{(n-1+\phi)^2\cdot\left[\lambda_s h'(\phi_0+n-1+\phi)\right]^2}{\lambda_b+\lambda_s h(\phi_0+n-1+\phi)} \mathrm{d}\phi \\
&= P^3\sum_{n=1}^{N_p}\int_{0}^{1} \frac{(n-1+\phi)^2\cdot\left[\lambda_s h'(\phi_0+\phi)\right]^2}{\lambda_b+\lambda_s h(\phi_0+\phi)} \mathrm{d}\phi. \qquad (4.32)
\end{aligned}
$$

By changing the integration variable in (4.32),

$$I_{22} = P^3\sum_{n=1}^{N_p}\int_{\phi_0}^{\phi_0+1} \frac{(n-1+\phi-\phi_0)^2\cdot\left[\lambda_s h'(\phi)\right]^2}{\lambda_b+\lambda_s h(\phi)} \mathrm{d}\phi. \qquad (4.33)$$

Because $N_p \gg 1$, the integral value of the term that contains $(n-1)^2$ is dominant. Therefore, I_{22} can be approximated by

$$I_{22} \approx P^3 \sum_{n=1}^{N_p} \int_{\phi_0}^{\phi_0+1} \frac{(n-1)^2 \cdot \left[\lambda_s h'(\phi)\right]^2}{\lambda_b + \lambda_s h(\phi)} \mathrm{d}\phi \tag{4.34}$$

which equals

$$I_{22} = P^3 \frac{(N_p-1)(N_p)(2N_p-1)}{6} \int_{\phi_0}^{\phi_0+1} \frac{\left[\lambda_s h'(\phi)\right]^2}{\lambda_b + \lambda_s h(\phi)} \mathrm{d}\phi \tag{4.35}$$

and can be approximated by

$$\begin{aligned} I_{22} &\approx P^3 \frac{N_p^3}{3} \int_{\phi_0}^{\phi_0+1} \frac{\left[\lambda_s h'(\phi)\right]^2}{\lambda_b + \lambda_s h(\phi)} \mathrm{d}\phi \\ &\approx \frac{T_{\mathrm{obs}}^3}{3} \int_0^1 \frac{\left[\lambda_s h'(\phi)\right]^2}{\lambda_b + \lambda_s h(\phi)} \mathrm{d}\phi. \end{aligned} \tag{4.36}$$

Substituting (4.23) and (4.24) into (4.7), changing the integration variable and splitting the integral, give rise to

$$\begin{aligned} I_{12} = I_{21} &= \int_{t_0}^{t_0+T_{\mathrm{obs}}} \frac{\frac{\partial}{\partial \phi_0}\lambda(t;\theta) \cdot \frac{\partial}{\partial f_o}\lambda(t;\theta)}{\lambda(t,\theta)} \mathrm{d}t \\ &= \int_{t_0}^{t_0+T_{\mathrm{obs}}} \frac{\left[\lambda_s h'(\phi_0+(t-t_0)f_o)\right]\left[\lambda_s (t-t_0) h'(\phi_0+(t-t_0)f_o)\right]}{\lambda_b + \lambda_s h(\phi_0+(t-t_0)f_o)} \mathrm{d}t \\ &= \int_0^{T_{\mathrm{obs}}} \frac{t \cdot \left[\lambda_s h'(\phi_0+f_o t)\right]^2}{\lambda_b + \lambda_s h(\phi_0+f_o t)} \mathrm{d}t \\ &= \sum_{n=1}^{N_p} \left(\int_{(n-1)P}^{nP} \frac{t \cdot \left[\lambda_s h'(\phi_0+f_o t)\right]^2}{\lambda_b + \lambda_s h(\phi_0+f_o t)} \mathrm{d}t \right) \\ &\quad + \int_{N_p P}^{N_p P + T_p} \frac{t \cdot \left[\lambda_s h'(\phi_0+f_o t)\right]^2}{\lambda_b + \lambda_s h(\phi_0+f_o t)} \mathrm{d}t \end{aligned} \tag{4.37}$$

As $N_p \gg 1$, the last integral in (4.37) can be neglected compared to the first term. Using the change of variable (4.31), (4.37) can be simplified as

$$
\begin{aligned}
I_{12} = I_{21} &\approx \sum_{n=1}^{N_p} \int_{(n-1)P}^{nP} \frac{t \cdot \left[\lambda_s h'(\phi_0 + f_o t)\right]^2}{\lambda_b + \lambda_s h(\phi_0 + f_o t)} \mathrm{d}t \\
&= \sum_{n=1}^{N_p} \int_0^1 \frac{(n-1+\phi)P \cdot \left[\lambda_s h'(\phi_0 + (n-1+\phi) f_o P)\right]^2}{\lambda_b + \lambda_s h(\phi_0 + (n-1+\phi) f_o P)} P \mathrm{d}\phi \\
&= P^2 \sum_{n=1}^{N_p} \int_0^1 \frac{(n-1+\phi) \left[\lambda_s h'(\phi_0 + n - 1 + \phi)\right]^2}{\lambda_b + \lambda_s h(\phi_0 + n - 1 + \phi)} \mathrm{d}\phi \\
&= P^2 \sum_{n=1}^{N_p} \int_0^1 \frac{(n-1+\phi) \left[\lambda_s h'(\phi_0 + \phi)\right]^2}{\lambda_b + \lambda_s h(\phi_0 + \phi)} \mathrm{d}\phi \\
&= P^2 \sum_{n=1}^{N_p} \int_{\phi_0}^{\phi_0+1} \frac{(n-1+\phi)[\lambda_s h'(\phi)]^2}{\lambda_b + \lambda_s h(\phi)} \mathrm{d}\phi.
\end{aligned} \tag{4.38}
$$

The term containing $(n-1)$ has the largest contribution in the summation (4.38), because $N_p \gg 1$. Therefore, (4.38) can be approximated by

$$
\begin{aligned}
I_{12} = I_{21} &\approx P^2 \sum_{n=1}^{N_p} \int_{\phi_0}^{\phi_0+1} \frac{(n-1)[\lambda_s h'(\phi)]^2}{\lambda_b + \lambda_s h(\phi)} \mathrm{d}\phi \\
&= P^2 \frac{(N_p - 1)N_p}{2} \int_{\phi_0}^{\phi_0+1} \frac{[\lambda_s h'(\phi)]^2}{\lambda_b + \lambda_s h(\phi)} \mathrm{d}\phi.
\end{aligned} \tag{4.39}
$$

Since the term containing N_p^2 is dominant, (4.39) can be approximated by

$$
I_{12} = I_{21} \approx P^2 \frac{N_p^2}{2} \int_{\phi_0}^{\phi_0+1} \frac{[\lambda_s h'(\phi)]^2}{\lambda_b + \lambda_s h(\phi)} \mathrm{d}\phi \tag{4.40}
$$

and using the approximation $T_{\mathrm{obs}} \approx N_p P$, it can be simplified to

$$
I_{12} = I_{21} \approx \frac{T_{\mathrm{obs}}^2}{2} \int_0^1 \frac{[\lambda_s h'(\phi)]^2}{\lambda_b + \lambda_s h(\phi)} \mathrm{d}\phi. \tag{4.41}
$$

From (4.28), (4.36), and (4.41), the Fisher matrix and its inverse can be found. These matrices are given in (4.10) and (4.12), respectively. ■

If the observed frequency, f_o, is known then the only unknown parameter to be estimated is ϕ_0. In this case, from (4.28), the CRLB for estimation of ϕ_0 is given by

$$\mathrm{CRLB}(\phi_0) = \frac{1}{T_{\mathrm{obs}} I_p}, \tag{4.42}$$

where I_p is given in (4.11).

The pulse delay estimate, $\hat{t}_{\mathrm{d}}$, can be determined from (4.3) as

$$\hat{t}_{\mathrm{d}} = \frac{\hat{\phi}_1 - \hat{\phi}_2}{\hat{f}_2}. \tag{4.43}$$

Hence, if velocities of the spacecraft are known, f_2 does not need to be estimated, and from (4.42) and (5.1), the CRLB for estimation of t_{d} is given by the following [44].

$$\mathrm{CRLB}(t_{\mathrm{d}}) = \frac{2}{f_2^2}\left(\frac{1}{T_{\mathrm{obs}} I_p}\right) \tag{4.44}$$

4.4 Discussion

Pulse delay estimator (4.43) is just a function of the velocity of spacecraft farther to the X-ray source. This may seem counter intuitive as one expects it to be a function of velocities of the both spacecraft. However, this is not the case. The pulsar signal is first detected on the closer vehicle. Then, assuming the clocks on both detectors are synchronized, since the signal is emitted by the X-ray source, the time duration it takes for the signal to be detected on the farther vehicle is just a function of its own velocity.

Equation (4.44) shows that the CRLB value for estimation of the pulse delay is inversely proportional to the observation time. In other words, to obtain more accurate pulse delay estimates, longer observation time is needed. Furthermore, since the CRLB is inversely proportional to I_p, which, in turn, is proportional to λ_{s}, making measurements on brighter pulsars results in more accurate pulse delay estimates. All of these analytical results are also consistent with intuition.

4.5 Numerical Examples

The CRLB for estimation of the relative distance is calculated for eight different X-ray pulsars. It is given by $\sigma = c\sqrt{\mathrm{CRLB}(t_{\mathrm{d}})}$, where c is the speed of light. The employed pulsars are presented in Table 4.1, and their profiles in two cycles are shown in Fig. 4.2. The spacecraft velocities are assumed to be $v_1 = 3$ km/s and $v_2 = 9$ km/s; and the initial phase observed at the first detector is $\phi_1 = 0.2$ cycle. The relative distance between the spacecraft is assumed to be 180 km. Hence, the pulse delay is $t_{\mathrm{d}} = 0.6$ ms.

Table 4.1 Accuracy of the X-ray pulsar measurements

Pulsar	Period (s)	$\min_{\lambda_b,\lambda_s} \sigma$ (m)
B0531+21	0.0335	2.6497E3
B0540−69	0.0504	4.4901E3
B0833−45	0.0893	1.8775E3
B1509−58	0.1502	2.8866E3
B1821−24	0.0031	188.68E0
B1937+21	0.0016	27.516E0
B1055−52	0.1971	1.1775E4
J0437−47	0.0057	461.08E0

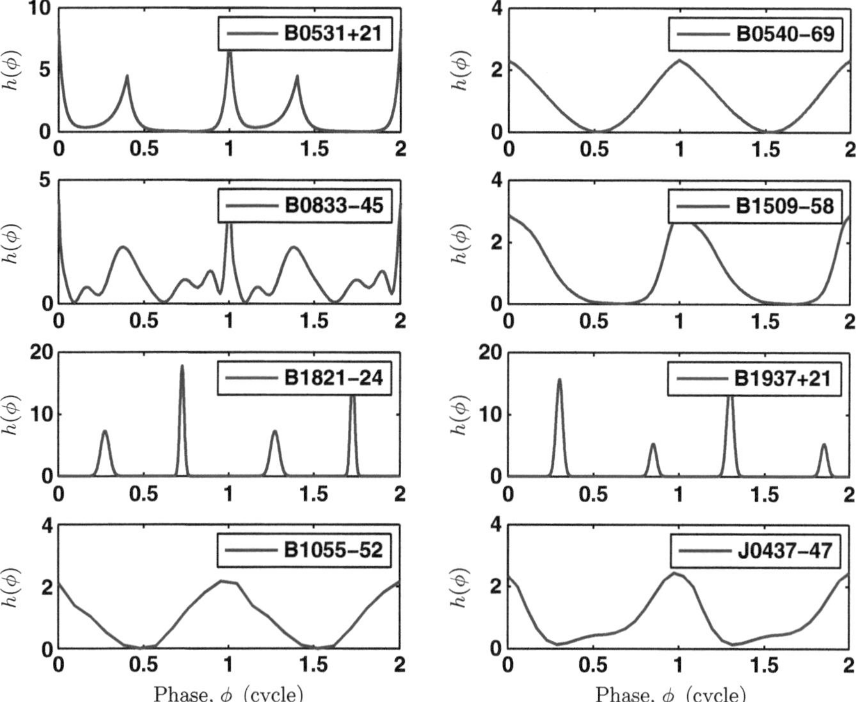

Fig. 4.2 Pulsars' profiles, $h(\phi)$, over two cycles

The observation time is assumed to be $T_{\text{obs}} = 100$ s, and σ is calculated for different values of λ_b and λ_s. The results are plotted in Fig. 4.3. As these figures show, the rate of change of the σ surface drops for large values of λ_s. Note that from (4.44), the σ value is inversely proportional to λ_s and the observation time, T_{obs}. Hence, to obtain smaller values of σ for relative distance estimation, either the observation time or λ_s must be increased. The minimum values of the σ surfaces are also given in Table 4.1. The minimum values change in a range of a few tens of meters to a few tens of kilometers.

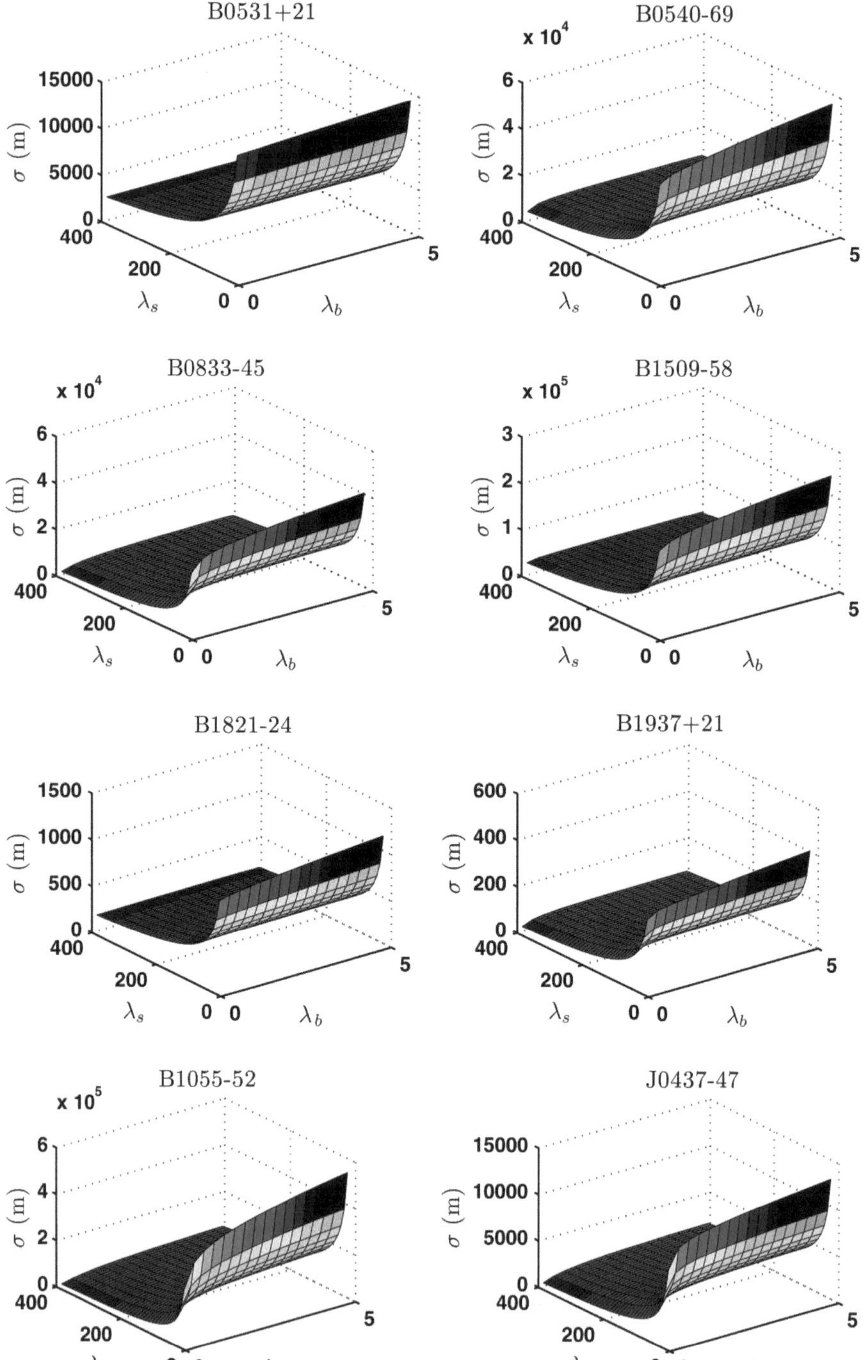

Fig. 4.3 The σ surface for $T_{\text{obs}} = 100\,\text{s}$

4.6 Summary

In this chapter, we formulate the pulse delay estimation problem. We show that the pulse delay is composed of the true time delay between the detected pulsar rate functions (which is due to the geometric range) and the differential time between the detectors' asynchronized clocks. We present the CRLB for estimation of the spacecraft absolute velocities and the initial phase of the pulsar intensity functions. We also present the CRLB for estimation of the pulse delay when the spacecraft velocities are known. Some numerical examples are given as well.

Chapter 5
Pulse Delay Estimation Using Epoch Folding

5.1 Introduction

In this chapter we offer our first approach for estimation of the pulse delay. The proposed approach is to retrieve the photon intensity functions on each detector via epoch folding to obtain the empirical rate functions, i.e., $\check{\lambda}_k(t)$. Then, the empirical intensities will be used for estimation of the initial phase on each detector. Finally, from (4.3), the pulse delay can be estimated using the estimates of ϕ_1 and ϕ_2,

$$\hat{t}_\mathrm{d} = \frac{\hat{\phi}_1 - \hat{\phi}_2}{f_2}. \tag{5.1}$$

Based on epoch folding, two different pulse delay estimators are presented in this chapter. Section 5.2 suggests using the correlation function between the empirical rate function and the true one to estimate the initial phase of the intensity function. In Sect. 5.3, a least-squares approach is proposed for estimation of the initial phase. Some clarifying remarks are given in Sect. 5.4. The effect of absolute velocity errors on the performance of the proposed pulse delay estimators is studied in Sect. 5.5. We study the numerical implementation of the estimators in Sect. 5.6. Finally, we provide several numerical examples to assess the performance of the proposed estimators in Sect. 5.7.

5.2 Cross Correlation Technique

Let T_b be the bin size, and $\check{\lambda}_j(t_i)$ be the empirical rate function on jth detector, where $0 \leq t_i \leq P$. The known rate function is

$$\lambda(t_i) = \lambda_\mathrm{b} + \lambda_\mathrm{s} h\big((t_i - t_0) f_\mathrm{o}\big). \tag{5.2}$$

Let the delayed rate function be

$$\lambda(t_i;\phi_j) = \lambda_\mathrm{b} + \lambda_\mathrm{s} h\big(\phi_j + (t_i - t_0) f_\mathrm{o}\big). \tag{5.3}$$

A.A. Emadzadeh and J.L. Speyer, *Navigation in Space by X-ray Pulsars*,
DOI 10.1007/978-1-4419-8017-5_5, © Springer Science+Business Media, LLC 2011

Therefore,

$$\breve{\lambda}_j(t_i) = \lambda(t_i;\phi_j) + \breve{n}_j(t_i), \quad j = 1,2, \tag{5.4}$$

where

$$E[\breve{n}_j(t_i)] = 0 \tag{5.5}$$

and

$$\begin{aligned} \mathrm{var}[\breve{n}_j(t_i)] &\triangleq \sigma_j^2(t_i) \\ &= \frac{N_\mathrm{b}}{T_\mathrm{obs}} \lambda_j(t_i;\phi_j). \end{aligned} \tag{5.6}$$

The initial phase ϕ_j can be estimated by identification of the maximum of the cross correlation (CC) function between $\lambda(t)$ and $\breve{\lambda}_j(t)$

$$\hat{\phi}_j = -\arg\max_{\psi\in(0,1)} R_j(\psi), \quad j = 1,2, \tag{5.7}$$

where

$$R_j(\psi) = \int_{-\infty}^{\infty} \lambda^*(t)\breve{\lambda}_j(t;\psi)\mathrm{d}t \tag{5.8}$$

is the CC function [44].

The asymptotic performance of this estimator is presented by the following theorem [44].

Theorem 5.1. *The pulse delay estimator obtained from (5.7) and (5.1), is unbiased for $T_\mathrm{obs} \gg 0$, and its asymptotic variance is given by,*

$$var[\hat{t}_\mathrm{d}] = \frac{2\int_0^1 (\lambda_\mathrm{b} + \lambda_\mathrm{s}h(\phi))\,[\lambda_\mathrm{s}h'(\phi)]^2\mathrm{d}\phi}{f_2^2 T_\mathrm{obs}\left(\int_0^1 [\lambda_\mathrm{s}h'(\phi)]^2\mathrm{d}\phi\right)^2}, \tag{5.9}$$

where $h'(\phi) = \partial h/\partial\phi$. This means that the estimator is consistent since its variance converges to zero as more TOA measurements are incorporated. Furthermore, it is not asymptotically efficient, meaning that its variance does not converge to the CLRB in finite time, i.e.,

$$var[\hat{t}_\mathrm{d}] > CRLB(t_\mathrm{d}) \quad for \quad T_\mathrm{obs} < \infty. \tag{5.10}$$

Proof. First, the performance of the cross correlation initial phase estimator on each detector is analyzed. For simplicity in notations, the subindex j of ϕ_j in (5.4) is dropped. Let

$$\begin{aligned} x_1(t) &= \lambda(t) \\ &= \lambda_\mathrm{b} + \lambda_\mathrm{s}h\big((t-t_0)f_\mathrm{o}\big) \end{aligned} \tag{5.11}$$

and

$$x_2(t) = \breve{\lambda}(t;\phi). \tag{5.12}$$

Assuming $\phi = -f_o D$, $x_2(t)$ can be rewritten as

$$\begin{aligned} x_2(t) &= \lambda_b + \lambda_s h\big((t - t_0)f_o + \phi\big) + \breve{n}(t) \\ &= \lambda_b + \lambda_s h\big((t - t_0 - D)f_o\big) + \breve{n}(t) \\ &= x_1(t - D) + \breve{n}(t). \end{aligned} \tag{5.13}$$

Let $\hat{D}$ be the estimate of D. As the signals $x_1(t)$ and $x_2(t)$ are band-limited, the cross correlation (5.8) has a derivative at $\hat{D}$, and it follows from (5.7) that

$$\dot{R}(\hat{D}) = 0. \tag{5.14}$$

By linearizing $\dot{R}(\hat{D})$ about the true time delay D, the following is obtained [53].

$$\dot{R}(D) + (\hat{D} - D)\ddot{R}(D) = 0 \tag{5.15}$$

Defining the delay estimation error

$$\delta = \hat{D} - D \tag{5.16}$$

from (5.16) and (5.15), it can be expressed as

$$\delta = -\frac{\dot{R}(D)}{\ddot{R}(D)}. \tag{5.17}$$

The signals $x_1(t)$ and $x_2(t)$ have the following spectra

$$S_1(f) = S_\lambda(f) \tag{5.18}$$

and,

$$S_2(f) = \begin{cases} S_\lambda(f)\mathrm{e}^{-j2\pi fD} + T_b \sum_k \breve{n}(k)\mathrm{e}^{-j2\pi fkT_b} & |f| < f_s/2 \\ 0 & |f| \geq f_s/2, \end{cases} \tag{5.19}$$

where $k \triangleq t_k$, $f_s = 1/T_b$, and $S_\lambda(f)$ is the spectrum of $\lambda(t)$, which equals

$$S_\lambda(f) = T_b \sum_k \lambda(k)\mathrm{e}^{-j2\pi fkT_b}. \tag{5.20}$$

The cross-correlation function is the inverse fourier transform of the cross power spectrum

$$R(\tau) = \int_{-\infty}^{\infty} S_1^*(f)S_2(f)\mathrm{e}^{j2\pi f\tau}df. \tag{5.21}$$

Therefore,

$$\begin{aligned}\dot{R}(\tau) &= j2\pi \int_{-\infty}^{\infty} f S_1^*(f) S_2(f) \mathrm{e}^{j2\pi f \tau} \mathrm{d}f \\ &= j2\pi \int_{-f_s/2}^{f_s/2} f S_\lambda^*(f) \left(S_\lambda(f) \mathrm{e}^{-j2\pi f D} + T_b \sum_k \breve{n}(k) \mathrm{e}^{-j2\pi f k T_b} \right) \mathrm{e}^{j2\pi f \tau} \mathrm{d}f.\end{aligned} \tag{5.22}$$

Assuming $\breve{n}(k)$ is small (or equivalently $T_{\mathrm{obs}} \gg 0$), (5.22) can be linearized about the true values of the samples as follows [53].

$$\dot{R}(\tau) \approx \sum_k \breve{n}(k) \frac{\partial \dot{R}(\tau)}{\partial \breve{n}(k)}, \tag{5.23}$$

where

$$\frac{\partial \dot{R}(\tau)}{\partial \breve{n}(k)} = j2\pi T_b \int_{-f_s/2}^{f_s/2} f S_\lambda^*(f) \mathrm{e}^{-j2\pi f k T_b} \mathrm{e}^{j2\pi f \tau} \mathrm{d}f. \tag{5.24}$$

Therefore,

$$\dot{R}(\tau) \approx j2\pi T_b \sum_k \breve{n}(k) \left(\int_{-\infty}^{\infty} f S_\lambda(f) \mathrm{e}^{j2\pi f (kT_b - \tau)} \mathrm{d}f \right)^* . \tag{5.25}$$

Hence,

$$\begin{aligned}\dot{R}(D) &\approx j2\pi T_b \sum_k \left\{ \breve{n}(k) \left(\int_{-\infty}^{\infty} f S_\lambda(f) e^{j2\pi f (kT_b - D)} \mathrm{d}f \right)^* \right\} \\ &= T_b \sum_k \breve{n}(k) \dot{\lambda}(k).\end{aligned} \tag{5.26}$$

From (5.21),

$$\begin{aligned}\ddot{R}(\tau) &= -4\pi^2 \int_{-\infty}^{\infty} f^2 S_1^*(f) S_2(f) \mathrm{e}^{j2\pi f \tau} \mathrm{d}f \\ &\approx -4\pi^2 \int_{-\infty}^{\infty} f^2 |S_\lambda(f)|^2 \mathrm{e}^{j2\pi f (\tau - D)} \mathrm{d}f.\end{aligned} \tag{5.27}$$

Then,

$$\begin{aligned}\ddot{R}(D) &\approx -4\pi^2 \int_{-\infty}^{\infty} f^2 |S_\lambda(f)|^2 \mathrm{d}f \\ &= T_b \sum_k \dot{\lambda}^2(k).\end{aligned} \tag{5.28}$$

Substituting (5.26) and (5.28) into (5.17), the estimation error is given by

$$\delta \approx \frac{\sum_k \breve{n}(k)\dot{\lambda}(k)}{\sum_k \dot{\lambda}^2(k)}. \tag{5.29}$$

Therefore, from (5.29)

$$E[\delta] = 0. \tag{5.30}$$

Since $\phi = -f_0 D$, this means that the phase estimator is unbiased as well,

$$E[\hat{\phi}] = \phi. \tag{5.31}$$

Furthermore,

$$\mathrm{var}[\delta] = \frac{\sum_k \mathrm{var}[\breve{n}(k)] \cdot \dot{\lambda}^2(k)}{\left(\sum_k \dot{\lambda}^2(k)\right)^2}, \tag{5.32}$$

where $\mathrm{var}[\breve{n}(k)]$ is given in (3.42). Therefore,

$$\mathrm{var}[\delta] = \frac{1}{T_b N_p} \cdot \frac{\sum_k \lambda(k)\dot{\lambda}^2(k)}{\left(\sum_k \dot{\lambda}^2(k)\right)^2}. \tag{5.33}$$

Let $T_b \to 0$ (or equivalently $N_b \to \infty$), then (5.33) becomes

$$\begin{aligned} \mathrm{var}[\delta] &\approx \frac{1}{T_b N_p} \cdot \frac{\frac{1}{T_b}\int_0^P \lambda(t)\dot{\lambda}^2(t)\mathrm{d}t}{\left(\frac{1}{T_b}\int_0^P \dot{\lambda}^2(t)\mathrm{d}t\right)^2} \\ &= \frac{1}{N_p} \cdot \frac{\int_0^P \lambda(t)\dot{\lambda}^2(t)\mathrm{d}t}{\left(\int_0^P \dot{\lambda}^2(t)\mathrm{d}t\right)^2}. \end{aligned} \tag{5.34}$$

The phase value varies in the range $0 \le \phi \le 1$, and the time span is $0 \le t \le P$. As a result, they are related as $\phi = t/P$. Therefore,

$$\mathrm{d}t = P\mathrm{d}\phi \tag{5.35}$$

and

$$\frac{\partial}{\partial t}\lambda(t) = \frac{\partial}{P\partial\phi}\lambda(\phi). \tag{5.36}$$

Thus, (5.34) can be simplified as,

$$\begin{aligned}
\mathrm{var}[\delta] &= \frac{1}{N_p}\cdot\frac{\dfrac{1}{P}\displaystyle\int_0^1 \lambda(\phi)\dot{\lambda}^2(\phi)\mathrm{d}\phi}{\left(\dfrac{1}{P}\displaystyle\int_0^1 \dot{\lambda}^2(\phi)\mathrm{d}\phi\right)^2} \\
&= \frac{P}{N_p}\cdot\frac{\displaystyle\int_0^1 \lambda(\phi)\dot{\lambda}^2(\phi)\mathrm{d}\phi}{\left(\displaystyle\int_0^1 \dot{\lambda}^2(\phi)\mathrm{d}\phi\right)^2} \\
&= \frac{f_\mathrm{o}^2}{T_\mathrm{obs}}\cdot\frac{\displaystyle\int_0^1 \lambda(\phi)\dot{\lambda}^2(\phi)\mathrm{d}\phi}{\left(\displaystyle\int_0^1 \dot{\lambda}^2(\phi)\mathrm{d}\phi\right)^2}.
\end{aligned} \tag{5.37}$$

Since $\hat{\phi} = -f_\mathrm{o}\hat{D}$,

$$\mathrm{var}[\hat{\phi}_j] = \frac{1}{T_\mathrm{obs}}\cdot\frac{\displaystyle\int_0^1 \lambda(\phi)\dot{\lambda}^2(\phi)\mathrm{d}\phi}{\left(\displaystyle\int_0^1 \dot{\lambda}^2(\phi)\mathrm{d}\phi\right)^2}. \tag{5.38}$$

Therefore, substituting $\lambda(\phi) = \lambda_\mathrm{b} + \lambda_\mathrm{s}h(\phi)$ in (5.38), and using (5.1), the variance of the time delay estimator is determined as given in (5.9).

To compare the performance against the CRLB it is shown that $\mathrm{var}[\hat{t}_\mathrm{d}]/\mathrm{CRLB}(t_\mathrm{d})$ is greater than one. Using (5.9) and (4.44), the ratio equals

$$\frac{\mathrm{var}[\hat{t}_\mathrm{d}]}{\mathrm{CRLB}(t_\mathrm{d})} = \frac{\displaystyle\int_0^1 g_1(\phi)\mathrm{d}\phi\cdot\int_0^1 g_3(\phi)\mathrm{d}\phi}{\left(\displaystyle\int_0^1 g_2(\phi)\mathrm{d}\phi\right)^2}, \tag{5.39}$$

where

$$g_1(\phi) \triangleq (\lambda_\mathrm{b} + \lambda_\mathrm{s}h(\phi))[\lambda_\mathrm{s}h'(\phi)]^2 \tag{5.40}$$

$$g_2(\phi) \triangleq [\lambda_\mathrm{s}h'(\phi)]^2 \tag{5.41}$$

$$g_3(\phi) \triangleq \frac{[\lambda_\mathrm{s}h'(\phi)]^2}{\lambda_\mathrm{b} + \lambda_\mathrm{s}h(\phi)}. \tag{5.42}$$

Note that from Cauchy–Schwartz inequality,

$$\int_0^1 g_1(\phi)\mathrm{d}\phi \cdot \int_0^1 g_3(\phi)\mathrm{d}\phi \geq \left(\int_0^1 \sqrt{g_1(\phi)g_3(\phi)}\mathrm{d}\phi\right)^2. \tag{5.43}$$

From (5.40), (5.41), and (5.42), it can easily be seen that

$$g_2(\phi) = \sqrt{g_1(\phi)g_3(\phi)}. \tag{5.44}$$

Therefore, (5.43) and (5.39) result in

$$\mathrm{var}[\hat{t}_\mathrm{d}] \geq \mathrm{CRLB}(t_\mathrm{d}). \tag{5.45}$$

The equality holds if and only if $g_3(\phi)$ is a multiple of $g_1(\phi)$, which implies

$$\lambda_\mathrm{b} + \lambda_\mathrm{s} h(\phi) \equiv \text{const}. \tag{5.46}$$

Since this is not the case for a general $h(\phi)$, equality does not hold, and the variance does not attain the CRLB asymptotically. ∎

5.3 Nonlinear Least Squares Technique

The initial phase can be estimated by fitting the empirical rate function, $\breve{\lambda}_j(t)$, to the true known rate function, $\lambda(t)$, via solving a nonlinear least squares (NLS) optimization problem [43]. The NLS cost function is defined based on minimizing the difference between the empirical rate function and the true one at the jth detector, i.e., $\breve{\lambda}_j(t) - \lambda(t;\phi_j)$. Hence, the objective function, $J(\phi_j)$, is defined as follows.

$$J(\phi_j) = \sum_{i=1}^{N_b} \left(\breve{\lambda}_j(t_i) - \lambda(t_i;\phi_j)\right)^2, \quad j = 1,2 \tag{5.47}$$

The unknown initial phase is estimated by minimizing the cost function (5.47), as

$$\hat{\phi}_j = \underset{\phi_j \in (0,1)}{\arg\min}\, J(\phi_j), \quad j = 1,2. \tag{5.48}$$

The following theorem summarizes the performance of the proposed pulse delay estimator [44].

Theorem 5.2. *The pulse delay estimator obtained from (5.48) and (5.1) is asymptotically unbiased, and its asymptotic variance is given by (5.9). That is, the NLS estimator's asymptotic performance is the same as the CC estimator.*

Proof. For simplicity in notations, the subindex j in (5.47) is dropped and the cost function is restated as,

$$J(\phi) = \left(\check{\mathbf{x}} - \mathbf{f}(\phi)\right)^{\mathrm{T}} \left(\check{\mathbf{x}} - \mathbf{f}(\phi)\right), \tag{5.49}$$

where

$$\check{\mathbf{x}} \triangleq \left(\check{\lambda}(t_1)\ \check{\lambda}(t_2)\ \cdots\ \check{\lambda}(t_{N_{\mathrm{b}}})\right)^{\mathrm{T}} \tag{5.50}$$

and

$$\mathbf{f}(\phi) \triangleq \left(\lambda(t_1;\phi)\ \lambda(t_2;\phi)\ \cdots\ \lambda(t_{N_{\mathrm{b}}};\phi)\right)^{\mathrm{T}}. \tag{5.51}$$

Now, assuming that the selected minimum is in the correct neighborhood (or equivalently the observation time is long enough), the mean and variance of the estimator $\hat{\phi}$ can be approximated by linearizing $\mathbf{f}(\phi)$ about the true value of phase, ϕ^o, as

$$\mathbf{f}(\phi) \approx \mathbf{f}(\phi^o) + (\phi - \phi^o)\mathbf{h}, \tag{5.52}$$

where

$$\mathbf{h} \triangleq \left.\frac{\partial \mathbf{f}(\phi)}{\partial \phi}\right|_{\phi=\phi^o}. \tag{5.53}$$

Therefore, (5.49) becomes

$$J(\phi) = \left(\check{\mathbf{x}} - \mathbf{f}(\phi^o) + \mathbf{h}\phi^o - \mathbf{h}\phi\right)^{\mathrm{T}} \left(\check{\mathbf{x}} - \mathbf{f}(\phi^o) + \mathbf{h}\phi^o - \mathbf{h}\phi\right). \tag{5.54}$$

Noting that $\check{\mathbf{x}} - \mathbf{f}(\phi^o) + \mathbf{h}\phi^o$ is fixed and ϕ is the variable, the linear least-squares solution to the cost function (5.54) is given by

$$\begin{aligned} \hat{\phi} &= k^{-1}\mathbf{h}^{\mathrm{T}}\left(\check{\mathbf{x}} - \mathbf{f}(\phi^o) + \mathbf{h}\phi^o\right) \\ &= \phi^o + r^{-1}\mathbf{h}^{\mathrm{T}}\left(\check{\mathbf{x}} - \mathbf{f}(\phi^o)\right), \end{aligned} \tag{5.55}$$

where

$$r \triangleq \mathbf{h}^{\mathrm{T}}\mathbf{h}. \tag{5.56}$$

The estimation error is defined as

$$e_\phi = \hat{\phi} - \phi^o. \tag{5.57}$$

Hence, from (5.55)

$$E\left[e_\phi\right] = r^{-1}\mathbf{h}^{\mathrm{T}}(\phi^o)E\left[\check{\mathbf{x}} - \mathbf{f}(\phi^o)\right] \tag{5.58}$$

and

$$\mathrm{var}\left[e_\phi\right] = r^{-1}\mathbf{h}^{\mathrm{T}}\mathrm{var}\left[\check{\mathbf{x}} - \mathbf{f}(\phi^o)\right]\mathbf{h}r^{-1}. \tag{5.59}$$

Note that from (3.40)

$$\check{\mathbf{x}} - \mathbf{f}(\phi^o) = \check{\mathbf{n}}, \tag{5.60}$$

where

$$\check{\mathbf{n}} \triangleq \left(\check{n}(t_1)\ \check{n}(t_2)\ \cdots\ \check{n}(t_{N_{\mathrm{b}}})\right)^{\mathrm{T}}. \tag{5.61}$$

Hence, employing (3.41) and (3.42), the asymptotic mean and variance of the estimation error, as $T_{\mathrm{obs}} \to \infty$, are

$$E[e_\phi] = 0 \tag{5.62}$$

and

$$\mathrm{var}[e_\phi] = \frac{N_\mathrm{b}}{T_{\mathrm{obs}}} r^{-1} \mathbf{h}^\mathrm{T} \mathbf{L} \mathbf{h} r^{-1}, \tag{5.63}$$

where

$$\mathbf{L} \triangleq \mathrm{diag}\big(\lambda(t_1;\phi^o), \lambda(t_2;\phi^o), \ldots, \lambda(t_{N_\mathrm{b}};\phi^o)\big). \tag{5.64}$$

This means that the NLS estimator $\hat{\phi}_0$ is asymptotically unbiased. By substituting the values of r, $\mathbf{h}$, and $\mathbf{L}$ into (5.63), the asymptotic variance of initial phase estimator is given by

$$\mathrm{var}[\hat{\phi}_j] = \frac{N_\mathrm{b}}{T_{\mathrm{obs}}} \cdot \frac{\sum_{i=1}^{N_\mathrm{b}} \lambda(t_i;\phi^o)\left(\lambda'(t_i;\phi^o)\right)^2}{\left(\sum_{i=1}^{N_\mathrm{b}} \left(\lambda'(t_i;\phi^o)\right)^2\right)^2}. \tag{5.65}$$

By letting $N_\mathrm{b} \to \infty$ and substituting the value of $\lambda(\cdot)$ from (3.27), (5.65) can be replaced by

$$\mathrm{var}[\hat{\phi}_j] = \frac{\int_0^1 (\lambda_\mathrm{b} + \lambda_\mathrm{s} h(\phi))\,[\lambda_\mathrm{s} h'(\phi)]^2 \mathrm{d}\phi}{T_{\mathrm{obs}} \left(\int_0^1 [\lambda_\mathrm{s} h'(\phi)]^2 \mathrm{d}\phi\right)^2}. \tag{5.66}$$

Therefore, from (5.1), the time delay estimator's asymptotic variance equals $2\,\mathrm{var}[\hat{\phi}_j]/f_2^2$, which is given in (5.9). ∎

5.4 Discussion

In [54], it is stated that if some regularity conditions are satisfied, the NLS estimator is consistent. It is easy to check that the mentioned conditions are met by the NLS estimator (5.48). Furthermore, it is mentioned that the NLS estimator is not in general asymptotically efficient, and this property depends on the underlying distribution of the noise. The significance of Theorem 5.2 is that it states (5.48) is indeed not an asymptotically efficient estimator. A meaningful measure to compare the estimator's performance to the CRLB is the *asymptotic relative efficiency* (ARE), which is the ratio between the estimator's asymptotic variance and the CRLB [55]. The ARE for the proposed NLS estimator is calculated and given in (5.39). From (5.39), it is clear that the LNS estimator's ARE is only a function of the pulsar profile, $h(\phi)$, and the photon constant rates, λ_b and λ_s.

Also note that the proposed NLS phase estimation approach is different from the one given in [56], which is typically used for radio pulsar timing. The method in [56] is based on minimizing the difference between the Fourier transforms of the empirical rate function and the true one.

5.5 Absolute Velocity Errors

The presented pulse delay estimators were analyzed based on accurate knowledge of the spacecraft absolute velocities. If there are absolute velocity errors, the proposed pulse delay estimation approach still works under certain conditions. If the velocity errors in the direction of the X-ray source are small enough, the resulted pulse delay errors can be neglected. This is a result of the upper bound presented in (3.72). If the velocity errors on each spacecraft are almost equal, $\Delta v_1 \approx \Delta v_2$, then from (3.66), $\Delta\phi_1 \approx \Delta\phi_2$. This results in the same deterioration of initial phase estimates on each detector. Hence, from (5.1), as the difference between the phase estimates is employed for estimation of the pulse delay, it remains asymptotically unbiased. However, the variance of estimation error becomes larger than the case where the absolute velocities are perfectly known due to two reasons [44]. One is the effect of the velocity error on f_2 in the denominator of (5.1). It is given by

$$\Delta\hat{t}_{\mathrm{d}} = -\frac{\Delta\hat{\phi}}{f_2^2}\Delta f_2, \tag{5.67}$$

where

$$\Delta\hat{\phi} = \hat{\phi}_1 - \hat{\phi}_2 \tag{5.68}$$

and

$$\Delta f_2 = f_{\mathrm{s}}\frac{\Delta v_2}{c}. \tag{5.69}$$

Therefore, the time delay estimate error due to this velocity error equals

$$\begin{aligned} \Delta\hat{t}_{\mathrm{d}} &= -\Delta\hat{\phi}\frac{\Delta v_2}{c f_2^2} f_{\mathrm{s}} \\ &\approx -\Delta\hat{\phi}\frac{\Delta v_2}{c f_2}. \end{aligned} \tag{5.70}$$

This causes $c\Delta\hat{t}_{\mathrm{d}}$ meters of error in position estimation

$$\begin{aligned} \delta\Delta x &= c\Delta\hat{t}_{\mathrm{d}} \\ &\approx -\Delta\hat{\phi}\frac{\Delta v_2}{f_2}. \end{aligned} \tag{5.71}$$

The second reason that the variance of the pulse delay estimator increases is that the variance of the initial phase estimates becomes larger. These estimation errors can be modeled by increasing the variance of the measurement noise in the Kalman filtering stage.

If the aforementioned assumptions about spacecraft velocities are not valid, the absolute velocities must be estimated as well as the initial phase values. We propose two solutions for this problem.

One approach is to simultaneously estimate f_k and ϕ_k. This is possible by using the TOA pdfs (6.1) to form two likelihood functions, and then solve the two-dimensional ML estimation problems. More details about this approach are discussed in Chap. 6.

In a different approach, we propose to employ the geometry of pulsars in the sky to estimate the spacecraft absolute velocities. For more details on this method, please see Chap. 7.

5.6 Computational Complexity Analysis

5.6.1 *Epoch Folding*

To use the CC or NLS initial phase estimators, the empirical rate function must be obtained through epoch folding. To obtain it, from (3.39), $N_b(N_p-1)$ additions, and $2N_b$ divisions are needed. Hence, the total number of calculations for epoch folding is $N_b(N_p+1)$ flops. This shows that the amount of necessary calculations for epoch folding is a linear function of N_p. In other words, it roughly linearly grows as the observation time becomes longer.

5.6.2 *CC Estimator*

As $\breve{\lambda}(t;\psi)$ is a discrete signal which is available at multiples of T_b s, the discrete cross-correlation function must be calculated. It is given by

$$R_D(\psi) = \frac{1}{N_b}\sum_{k=1}^{N_b} x_1(kT_b)x_2(kT_b;\psi), \tag{5.72}$$

where $x_1(t)$ and $x_2(t)$ are defined in (5.11) and (5.12), respectively [57].

The cross-correlation function can be computed in the time domain using (5.72), or in the Fourier domain

$$R_D(\psi) = \mathscr{F}^{-1}\{X_1(f_k)X_2^*(f_k)\}, \tag{5.73}$$

where $X_1(f_k)$ and $X_2(f_k)$ are the discrete Fourier transforms of $x_1(kT_b)$ and $x_2(kT_b)$.

If the resolution of the initial phase estimate is needed to be finer than $1/N_b$, it is necessary to interpolate the cross-correlation function. One approach is to approximate the correlation function by a convex parabola in the neighborhood of its maximum, as

$$R_D(\psi) = a\psi^2 + b\psi + c, \tag{5.74}$$

where a, b, and c are parameters fitting the measured correlation. Using this approximation, the continuous time-delay estimation can be performed by finding the apex of the parabola,

$$\hat{D}_s = -\frac{b}{2a}. \tag{5.75}$$

Hence, to obtain the continuous initial phase estimate, two steps are needed:

- Locate the index k_m of the maximizer delay $R_D(k_m T_s)$, where $T_s = 1/N_b$, to find the coarse estimate.
- To find the subsample estimate, use the maximum cross-correlation delay $R_D(k_m T_s)$, and two of its adjacent values $R_D(k_m T_s - T_s)$ and $R_D(k_m T_s + T_s)$ for a parabolic approximation, and add it to the coarse estimate,

$$\hat{\phi} = k_m T_s - \frac{1}{2} \cdot \frac{R_D(k_m T_s + T_s) - R_D(k_m T_s - T_s)}{R_D(k_m T_s + T_s) - 2R_D(k_m T_s) + R_D(k_m T_s - T_s)}. \tag{5.76}$$

Note that due to the parabolic approximation, the phase estimate (5.76) becomes biased [57].

Using (5.72), to obtain the discrete cross-correlation function, for a fixed ψ, N_b multiplications, $N_b - 1$ additions, and 1 division are performed. Since the search domain to maximize $R_D(\psi)$ is $(0,1)$ cycle, which is divided into N_g grids, a total number of $2N_g N_b$ flops must be performed to obtain the initial phase estimate.

From (5.76), we can see that three additions and two multiplications are needed for interpolation. It is also clear that the interpolation is performed only to find the maximizer point. Hence, the number of calculation flops is not a function of the observation time, T_{obs}.

In summary, to obtain the CC estimator, first epoch folding is performed which needs $N_b(N_p + 1)$ operation flops. Then the search interval $(0,1)$ cycle is divided into N_g grids, and the correlation function is obtained for each grid. This imposes $2N_b N_g$ operations.

If the correlation function is obtained using the Fourier transform from (5.73), first the Fourier transform of each signal must be found. Then N_b multiplications are performed and using inverse Fourier transform on the resulted signal, the correlation function is determined. As fast Fourier transform techniques are widely available, this approach is expected to be faster than the time domain method.

5.6.3 NLS Estimator

To construct the NLS cost function, using (5.47), N_b subtractions, N_b multiplications, and N_b additions must be performed for a fixed ϕ_0. Hence, the total number of floating-point operations is $3N_b$. This means that the computational cost to construct the NLS cost function is not a function of the observation time.

Although for any given ϕ_0, $\lambda(t_i, \phi_0)$ is known, the mathematical equation of the rate function $\lambda(t; \phi_0)$ is not usually available in practice. Hence, the gradient of the cost function, if needed, must be calculated numerically. Furthermore, the cost function is not convex in general. This means that it usually has multiple minima. Hence, to avoid getting trapped in local extrema, and since (5.48) is a scalar optimization problem, employing a one-dimensional grid search over the interval $(0,1)$ cycle is necessary to solve it. In other words, first, the epoch folding procedure, with a cost of $N_b(N_p+1)$ flops, is performed. Then, the search interval $(0,1)$ is divided into N_g grids. For each grid, the cost function (5.47) is calculated, and the minimizer grid is found. This imposes a total floating-point operations of $3N_gN_b$ flops.

5.7 Numerical Examples

First, performance of the initial phase estimators on one detector is examined. Then, a relative navigation scenario is presented and the pulse delay estimators are simulated.

5.7.1 Initial Phase Estimators

The Crab pulsar (PSR B0531+21) is chosen as the X-ray source. The constant arrival rates are chosen to be $\lambda_b = 5$ and $\lambda_s = 15$ ph/s. The spacecraft velocity on the direction vector pointing to the pulsar is assumed to be $v = 3$ km/s; and the initial phase observed at the detector is $\phi_1 = 0.2$ cycle. A Monte Carlo simulation was performed in which 500 independent realization of photon TOAs were processed by the phase estimators.

To evaluate each estimator's performance, the empirical rate function is derived for each observation time, and the initial phase is estimated by optimizing the cost functions given in (5.8) and (5.47). The optimization is done using a grid search algorithm over $[0,1)$ cycle. The interval is divided into 1,024 grids and the cost functions are calculated for each one. Note that the magnitude of the estimation error is calculated modulo one cycle, i.e.,

$$|e| = \min\{\text{mod}\,(\phi_0 - \hat{\phi}_0, 1), \quad \text{mod}\,(\hat{\phi}_0 - \phi_0, 1)\}. \tag{5.77}$$

For instance, if $\phi_0 = 0.1$ and $\hat{\phi}_0 = 0.9$, the error is 0.2 cycle, and not 0.8 cycle.

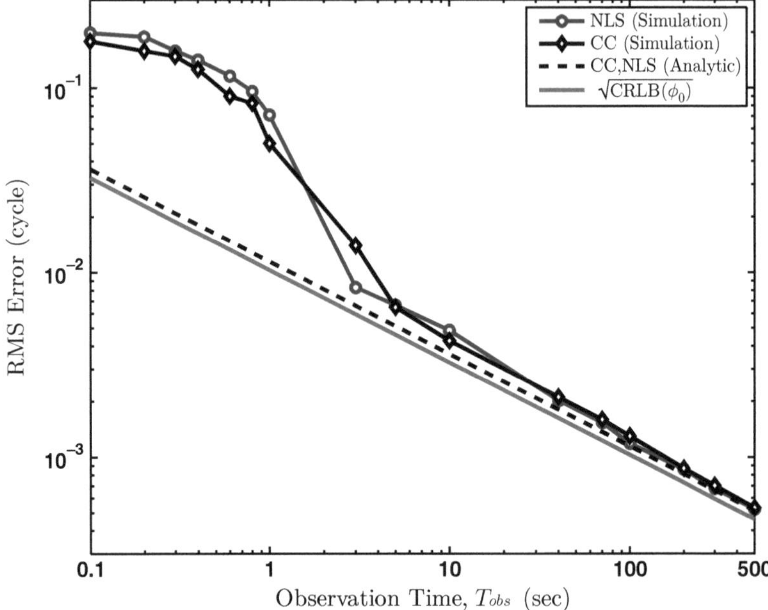

Fig. 5.1 RMS error for the CC and NLS phase estimators

In Fig. 5.1, the root mean square (RMS) of the CC and NLS estimation errors for each observation time are plotted against the square root of the $\mathrm{CRLB}(\phi_0)$ and the analytical value, presented in (4.42) and (5.66), respectively. The mean square error (MSE) of the estimator which measures the average mean squared deviation of the estimator from the true value, is defined as follows.

$$\mathrm{MSE}(\hat{\phi}_0) = E[(\hat{\phi}_0 - \phi_0)^2] \tag{5.78}$$

The MSE can be rewritten as

$$\begin{aligned}\mathrm{MSE}(\hat{\phi}_0) &= E\left\{\left[\left(\hat{\phi}_0 - E(\hat{\phi}_0)\right) + \left(E(\hat{\phi}_0) - \phi_0\right)\right]^2\right\} \\ &= \mathrm{var}(\hat{\phi}_0) + \left[E(\hat{\phi}_0) - \phi_0\right]^2 \\ &= \mathrm{var}(\hat{\phi}_0) + b^2(\phi_0),\end{aligned} \tag{5.79}$$

where

$$b(\phi_0) = E(\hat{\phi}_0) - \phi_0 \tag{5.80}$$

is the bias of the estimator. This shows that the MSE is composed of errors due to the variance of the estimator as well as the bias.

The RMS plots in Fig. 5.1 show that as time goes on, the variance of estimation error approaches zero, implying that both estimators are consistent. The difference between the RMS error and the CRLB is determined by the difference between (5.66) and (4.42). It is in general a function of the pulsar rate function, but it is always inversely proportional to the observation time. As the observation time is reduced, there is a threshold point at which the RMS error starts to deviate from the CRLB value. The reason is that as the observation time goes below this threshold, the realization of TOAs does not represent the rate function precisely. Therefore, they result in a distorted cost function whose extremum happens at a farther point to the true parameter, ϕ^o. As a result, the estimator becomes biased, and as (5.79) shows, the estimation error's variance starts to depart from the CRLB.

The CC cost function (5.8), obtained by Monte Carlo simulation over 500 realizations of TOAs, is plotted in Fig. 5.2 for different observation times. The plots show that for short observation times, the CC function has a smaller peak, and as the observation time increases and more TOAs are measured, the peak value becomes larger.

The cost function (5.47) is also plotted in Fig. 5.3 for different observation times. The plots show that as the observation time increases, the cost function becomes smaller.

Figures 5.2 and 5.3 also show that each of the cost functions has multiple extrema. Hence, they verify that, to avoid getting trapped in local extrema, adopting

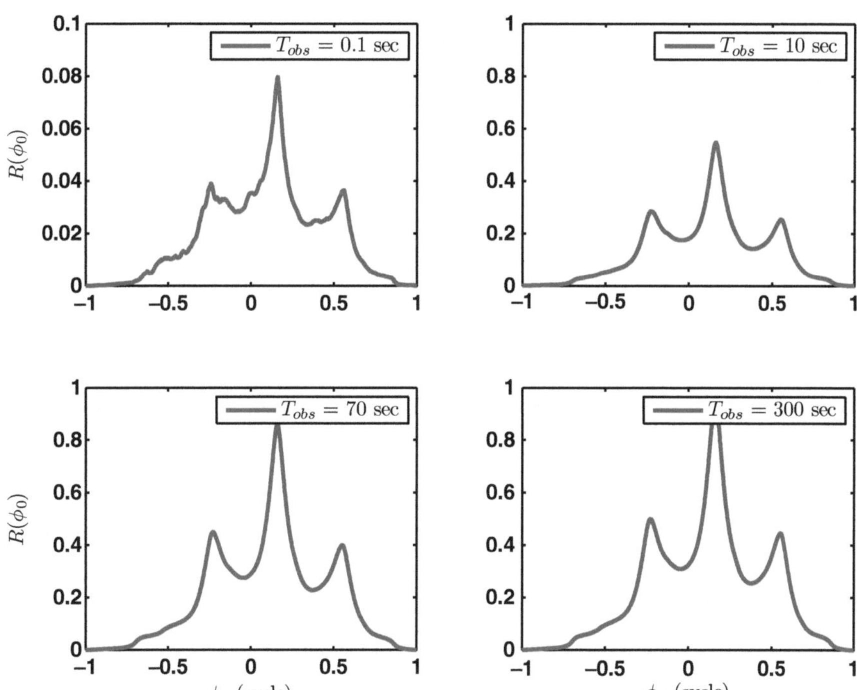

Fig. 5.2 The CC cost function

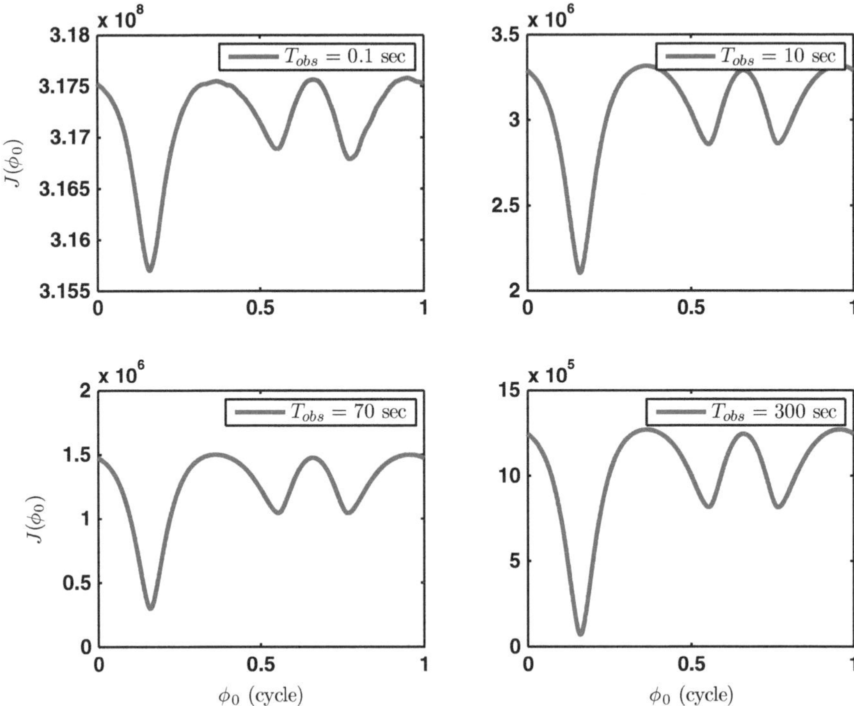

Fig. 5.3 The NLS cost function

a grid search optimization approach is necessary. As the search is performed over the whole interval $(0,1)$ cycle, the algorithm initialization issues do not arise.

5.7.2 Pulse Delay Estimators

Now we simulate the pulse delay estimators. The spacecraft velocities on the direction vector pointing to the pulsar are assumed to be $v_1 = 3$ km/s and $v_2 = 9$ km/s; and the initial phase observed at the first detector is $\phi_1 = 0.2$ cycle. The Doppler frequencies depend on v_1 and v_2, which are scalars. Hence, we do not need the spacecraft velocity vectors for this simulation. The relative distance between the spacecraft is assumed to be $\Delta d = 180$ km. Hence, the pulse delay to be estimated is $t_\mathrm{d} = 0.6$ ms. Since t_d is less than the pulsar period, there is no phase ambiguity.

An observation time of $T_\mathrm{obs} = 100$ s is selected to verify the epoch folding results. For each observation time, photon time tags are folded back into one single pulsar cycle which is divided into $N_\mathrm{b} = 1{,}024$ bins, and the rate functions on each detector are derived using (3.39). The empirical rate functions along with the true ones are plotted in Figs. 5.4 and 5.5. As the plots show, the empirical rate functions

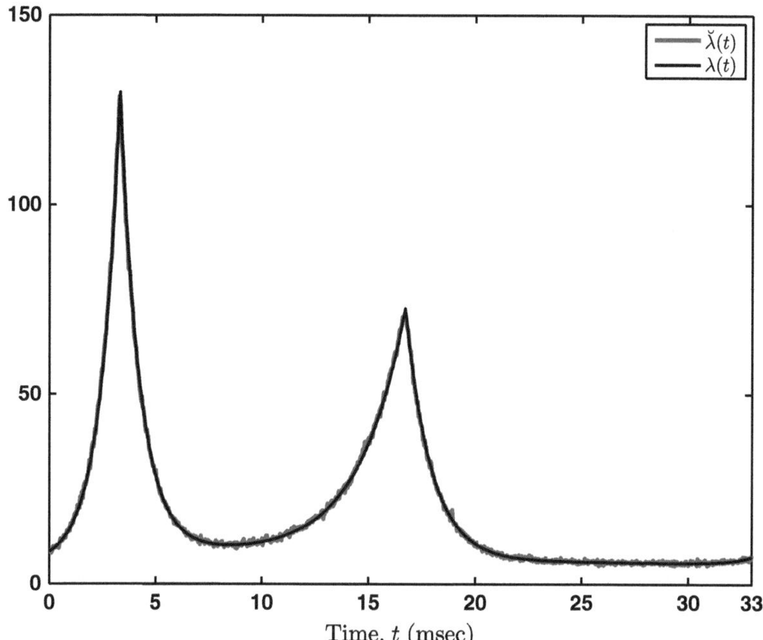

Fig. 5.4 $\breve{\lambda}_1(t)$ for $T_{\text{obs}} = 100\,\text{s}$

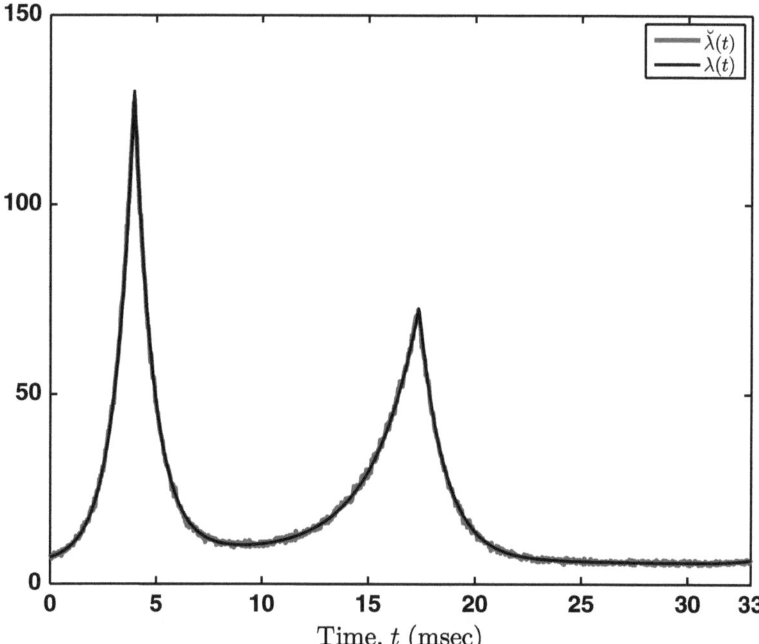

Fig. 5.5 $\breve{\lambda}_2(t)$ for $T_{\text{obs}} = 100\,\text{s}$

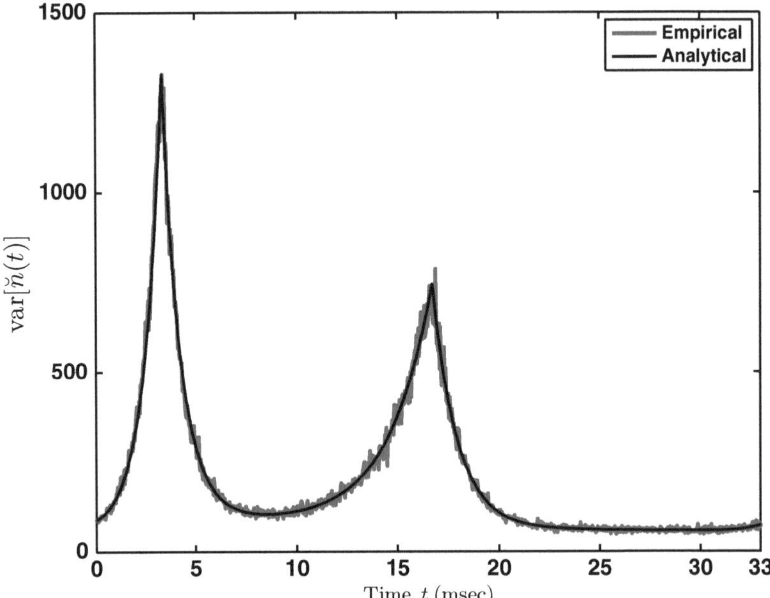

Fig. 5.6 The variance of $\breve{n}_1(t)$ for $T_{\text{obs}} = 100\,\text{s}$

converge to the true functions. In Figs. 5.6 and 5.7, the variance of $\breve{n}(t)$, obtained via simulation, is plotted and compared to the analytical variance for each detector.

The effect of imprecise absolute velocity data on epoch folding is studied as well. The errors are intentionally chosen to be big: $\Delta v_1 = 1{,}000$ m/s and $\Delta v_2 = 1{,}250$ m/s. For the selected parameters, the upper bound (3.72) roughly equals 10 m/s. Since the velocity errors are bigger than the bound, the empirical rate functions are expected to be deteriorated. Figures 5.8 and 5.9 verify this fact. Comparing them to the true rate functions shows that they are shifted, and their peaks are flattened. The same phenomenon happens for the noise variance which can be seen in Figs. 5.10 and 5.11.

Figures 5.12 and 5.13 show the RMS error of the proposed pulse delay estimators, obtained through simulation, against the CRLB and the analytical value, calculated as in (4.44) and (5.9). The mean square error (MSE) of the estimator is defined as

$$\text{MSE}(\hat{t}_\text{d}) = E[(\hat{t}_\text{d} - t_\text{d})^2] \tag{5.81}$$

and the cost functions are optimized using a grid search approach in the domain of $(0,1)$ cycle.

As expected, the RMS plots show that as time goes on, the variance of each pulse delay estimator approaches zero. Also, as the observation time is reduced, there is a threshold point at which the estimation error variance starts to deviate from the CRLB. It is due to the fact that the phase estimators become biased when there are not enough photon TOAs to process.

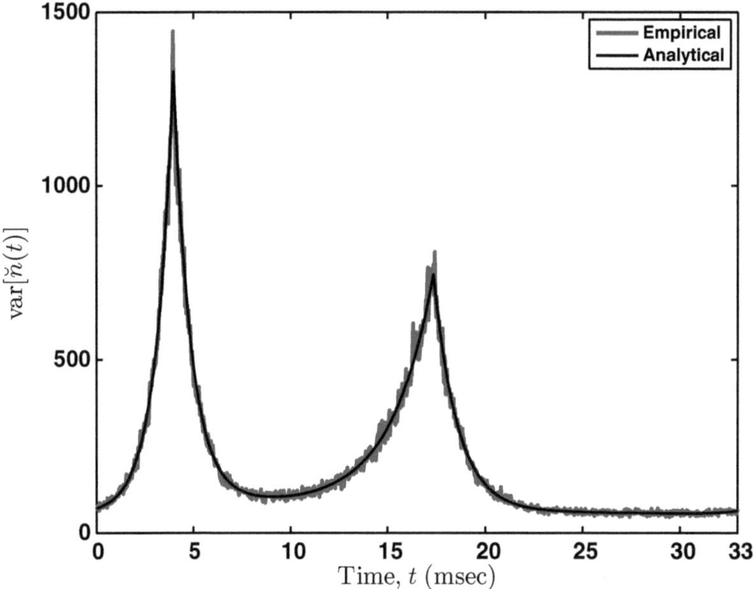

Fig. 5.7 The variance of $\breve{n}_2(t)$ for $T_{\mathrm{obs}} = 100\,\mathrm{s}$

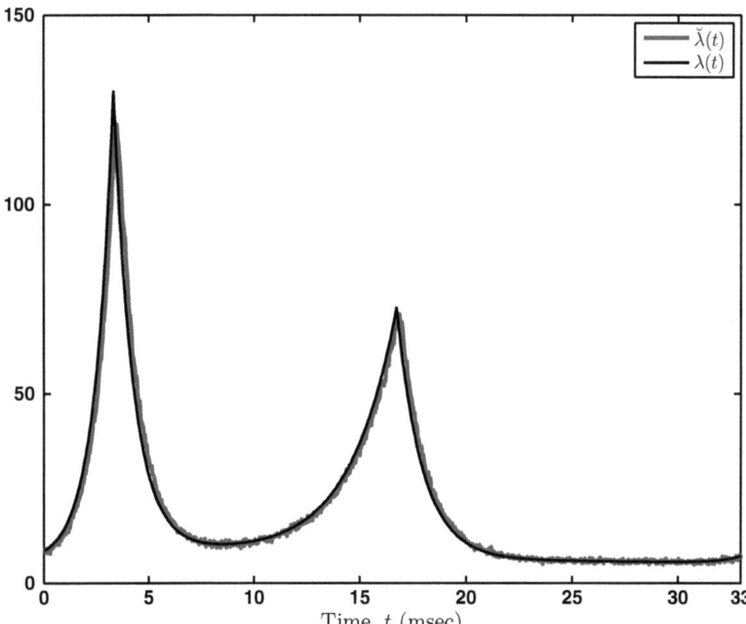

Fig. 5.8 $\breve{\lambda}_1(t)$ when $\Delta v_1 = 1{,}000\,\mathrm{m/s}$

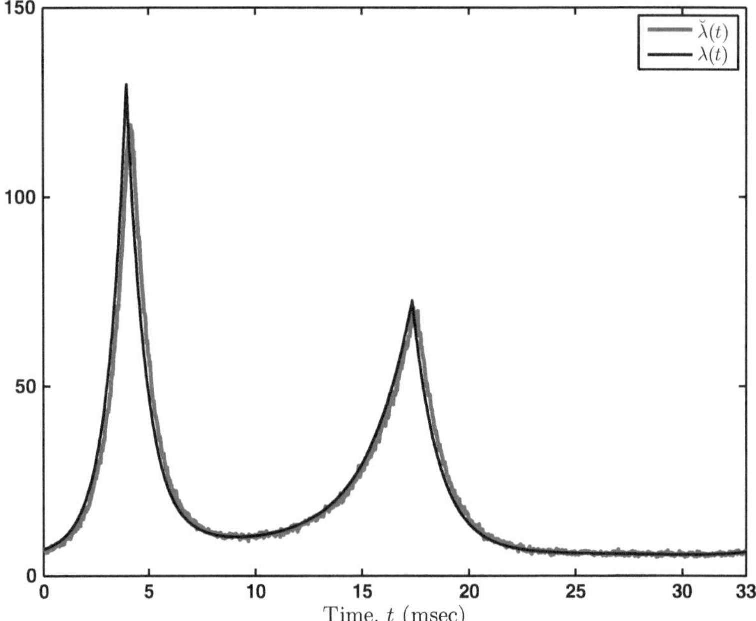

Fig. 5.9 $\breve{\lambda}_2(t)$ when $\Delta v_2 = 1{,}250$ m/s

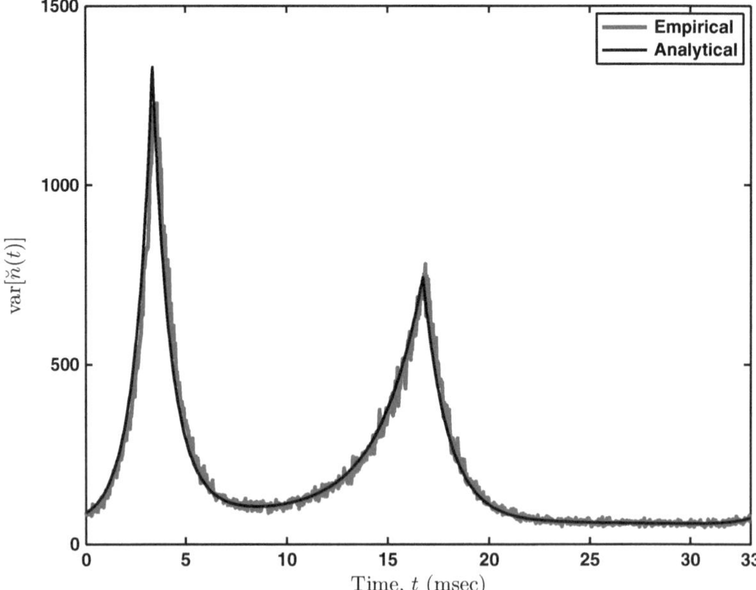

Fig. 5.10 var$[\breve{n}(t)]$ when $\Delta v_1 = 1{,}000$ m/s

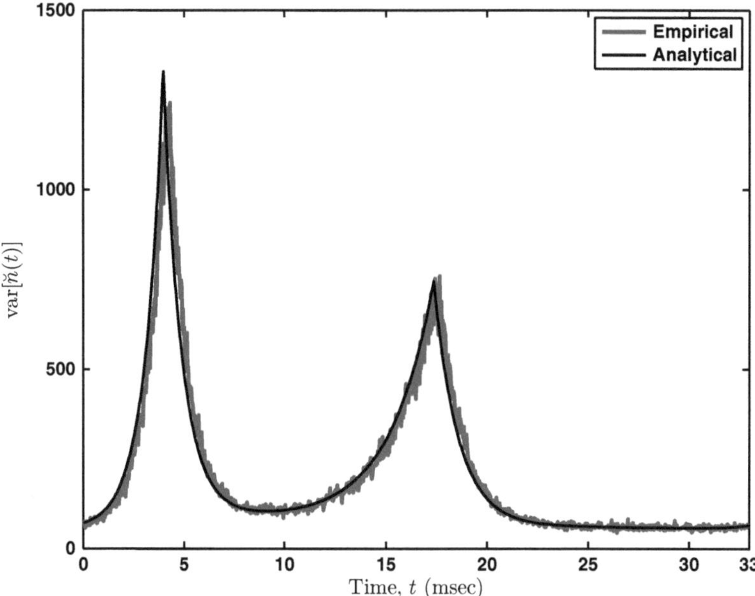

Fig. 5.11 var[$\breve{n}(t)$] when $\Delta v_2 = 1{,}250$ m/s

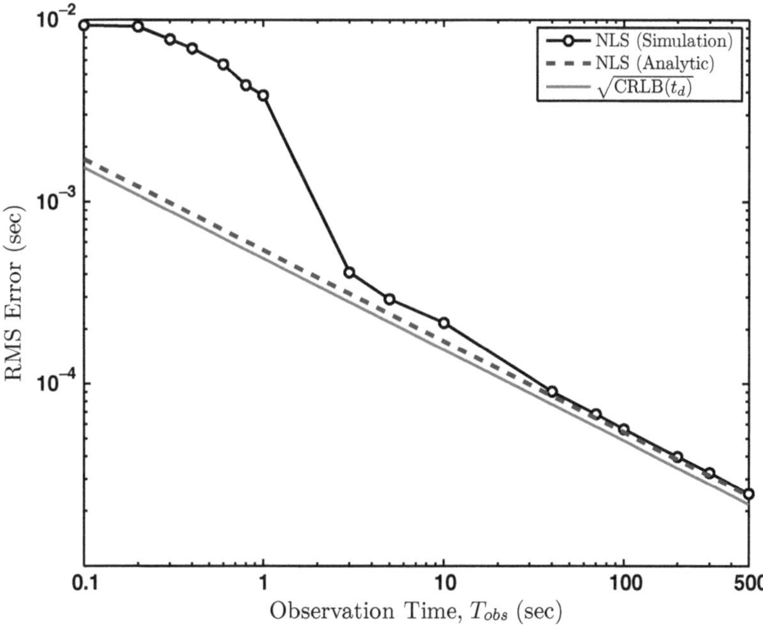

Fig. 5.12 The RMS error of the NLS-based estimator

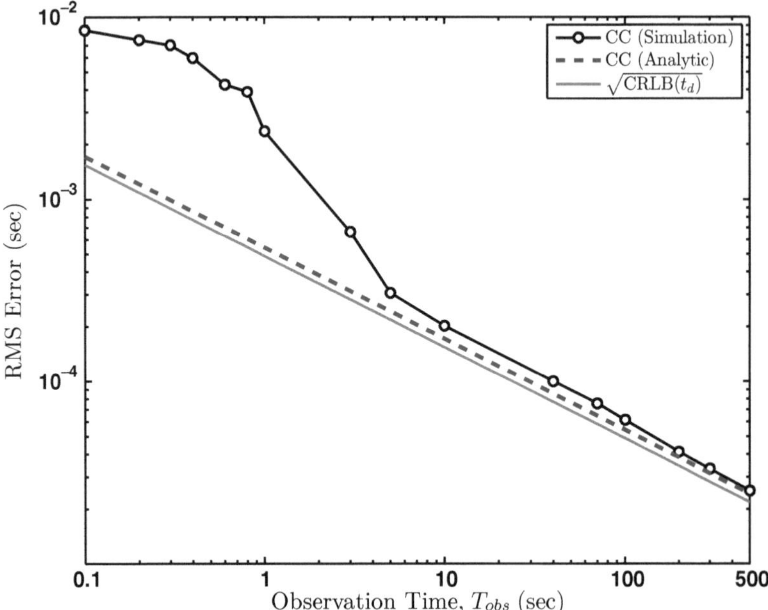

Fig. 5.13 The RMS error of the CC-based estimator

Note that we performed this simulation to verify that the analytical RMS values match the numerical results. Different choices of λ_b, λ_s, v_1, v_2, ϕ_1, and Δd, only change numerical values of CRLB, analytical RMS errors, and simulation RMS errors.

Finally, the performance of the pulse delay estimators is studies when the spacecraft velocity errors are $\Delta v_1 = 1{,}000$ m/s and $\Delta v_2 = 1{,}250$ m/s. The RMS error plots are presented in Figs. 5.14 and 5.15. As the plots show, the asymptotic performance of the estimators degrades compared to the case where the absolute velocities are perfectly known.

5.8 Summary

We explain how to use the epoch folding procedure for estimation of the pulse delay. We present two different initial phase estimators which both work based on epoch folding. The key idea in employing epoch folding for X-ray pulsar-based navigation is to recover the photon intensity functions on each detector and estimate their initial phase. One estimator works based on maximizing the cross correlation between the empirical rate function and the true one. The other one is obtained by minimizing the difference between the rate functions via solving a least-squares problem. We analyze theses estimators and show that they are not asymptotically efficient but they

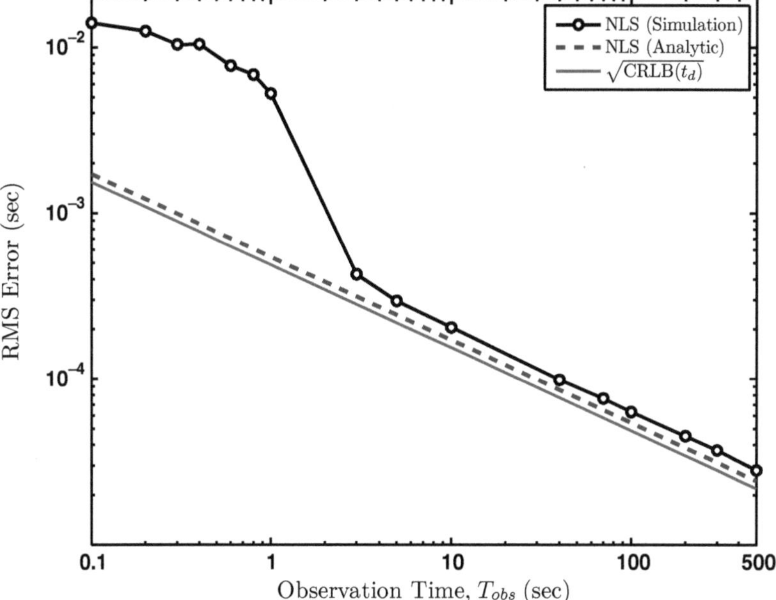

Fig. 5.14 The RMS error of the NLS-based estimator in presence of absolute velocity errors

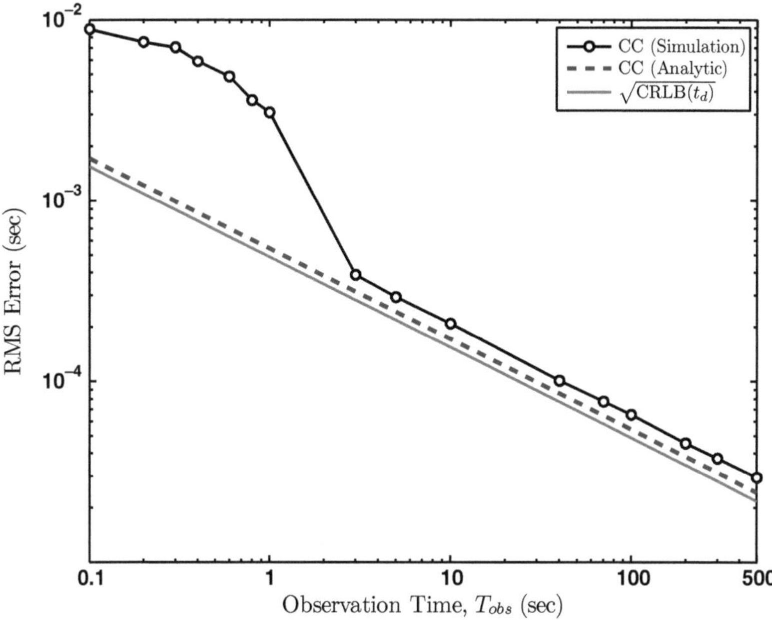

Fig. 5.15 The RMS error of the CC-based estimator in presence of absolute velocity errors

are consistent. We investigate the numerical implementation aspects by analyzing the number of floating point operations. Further numerical examples on computational complexity of the estimators will be presented in Chap. 6. We explain that the proposed pulse delay estimation technique works not only when the spacecraft velocities are perfectly known but also under some conditions where the velocity data is not perfect. By offering different simulation scenarios, we also examine the analytical results via numerical examples.

Chapter 6
Pulse Delay Estimation via Direct Use of TOAs

6.1 Introduction

An important point regarding the pulse delay estimation techniques introduced in Chap. 5 is that they need to employ the epoch folding procedure which needs the exact knowledge of the spacecraft velocities. Furthermore, it was shown that the proposed epoch folding-based estimators are not asymptotically efficient. To address these problems, we propose another method in this chapter. This approach is based on direct use of the measured TOA sets, $\{t_i^{(1)}\}_{i=1}^{M_1}$ and $\{t_i^{(2)}\}_{i=1}^{M_2}$. Because the measured TOAs are being used directly, it is not necessary for the estimators to have access to the velocity data. Nonetheless, the effect of imprecise velocity data on the performance of the pulse delay estimator is studied. Computational complexity study is also performed.

We formulate an ML problem in Sect. 6.2 for estimation of the Doppler frequencies and the initial phase values. Some remarks on numerical implementation of the proposed estimator are offered in Sect. 6.3. The ML computational complexity analysis is presented in Sect. 6.4. All computational complexity analysis results are summarized in Sect. 6.5. The effect of absolute velocity errors on the pulse delay estimator's performance is studied in Sect. 6.6. Numerical simulations are provided in Sect. 6.7.

6.2 Maximum-Likelihood Estimator

Employing the pdfs associated with the detected time tags, an ML estimation problem can be formulated to estimate ϕ_k and f_k. According to (3.10), the pdfs associated with each set of time tags are given by,

$$p\left(\{t_i^{(k)}\}_{i=1}^{M_k};\phi_k,f_k\right) = \mathrm{e}^{-\Lambda(\phi_k,f_k)}\prod_{i=1}^{M_k}\lambda_k(t_i^{(k)};\phi_k,f_k), \tag{6.1}$$

A.A. Emadzadeh and J.L. Speyer, *Navigation in Space by X-ray Pulsars*,
DOI 10.1007/978-1-4419-8017-5_6,

where

$$\Lambda(\phi_k, f_k) \triangleq \int_{t_0}^{t_f} \lambda_k(t; \phi_k, f_k)\mathrm{d}t, \quad k = 1, 2. \tag{6.2}$$

Recognizing the pdfs given in (6.1) as likelihood functions, the maximum-likelihood estimators (MLEs) are provided by maximizing each likelihood function with respect to the unknown parameters. Equivalently, the natural logarithm of the likelihood function or the log-likelihood function (LLF) can be maximized.

$$\mathrm{LLF}(\phi_k, f_k) = \sum_{i=1}^{M_k} \ln\left(\lambda_k(t_i; \phi_k, f_k)\right) - \Lambda(\phi_k, f_k), \quad k = 1, 2. \tag{6.3}$$

If the observation time is long enough compared with the pulsar period then $\Lambda(\phi_k, f_k)$ in (6.3) shows a minimal dependence on the parameters ϕ_k and f_k, that is, $\partial\Lambda/\partial\phi_k = 0$, and $\partial\Lambda/\partial f_k = 0$. The reason is as follows. Let the observation time contain N_p pulsar cycles, $T_{obs} = N_p P + T_p$, where $0 \leq T_p < 1$. Using periodicity of $\lambda(t; \phi_k, f_k)$, (6.2) can be res-stated as follows.

$$\Lambda(\phi_k, f_k) = \int_{t_0}^{t_0+T_p} \lambda(t; \phi_k, f_k)dt + N_p \int_0^P \lambda(t; \phi_k, f_k)dt. \tag{6.4}$$

Since $\lambda(\cdot) \geq 0$ is a periodic function, its integral over one period is not a function of the initial phase and the Doppler frequency. As a result, the second term on the right hand side of (6.4) is independent of ϕ_k and f_k, while the first term is a function of ϕ_k and f_k. Hence, if $N_p \gg 1$, then the second term is strongly dominant, and the value of $\Lambda(\phi_k, f_k)$ will not be sensitive to ϕ_k and f_k. In other words $\partial\Lambda(\phi_k)/\partial\phi_k \approx 0$ and $\partial\Lambda(f_k)/\partial f_k \approx 0$. Therefore, it can be dropped from the objective function, and the likelihood function, denoted by $\Psi(\phi_k, f_k)$, becomes,

$$\Psi(\phi_k, f_k) = \sum_{i=1}^{M_k} \ln\left(\lambda_k(t_i; \phi_k, f_k)\right). \tag{6.5}$$

The unknown parameters ϕ_k and f_k can be determined by solving the following optimization problems,

$$(\hat{\phi}_k, \hat{f}_k) = \underset{\phi_k, f_k}{\arg\max}\ \Psi(\phi_k, f_k), \quad k = 1, 2 \tag{6.6}$$

and the pulse delay may be estimated from (4.43).

The statistical properties of the MLEs (6.6) are summarized in the following theorem [51].

Theorem 6.1. *The MLE determined from (6.6) is unbiased for* $T_{\mathrm{obs}} \gg 0$, *and attains the CRLB, presented in (4.12). In other words, it is* asymptotically efficient.

Proof. To investigate the properties of the MLE, first note the following theorem [50].

Theorem 6.2. *If the pdf $p(x;\theta)$ of the data x satisfies the regularity conditions given in (4.5) then the MLE of the parameter θ is asymptotically unbiased and asymptotically attains the CRLB. It is therefore* asymptotically efficient.

Let θ be the unknown parameter to be estimated. To apply Theorem 6.2, it must be shown that the regularity condition (4.5) is satisfied, where $\mathbf{x} = \{t_i\}_{i=1}^{M}$. From (3.10),

$$\ln p(\mathbf{x};\theta) = -\Lambda(\theta) + \sum_{i=1}^{M} \ln\lambda(t_i;\theta). \tag{6.7}$$

Therefore,

$$\frac{\partial}{\partial\theta}\ln p(\mathbf{x};\theta) = -\frac{\partial}{\partial\theta}\Lambda(\theta) + \sum_{i=1}^{M} \frac{1}{\lambda(t_i;\theta)}\cdot\frac{\partial}{\partial\theta}\lambda(t_i;\theta). \tag{6.8}$$

Because the observation time is assumed to be long enough, the first term in (6.8) vanishes and,

$$E\left[\frac{\partial}{\partial\theta}\ln p(\mathbf{x};\theta)\right] = \sum_{i=1}^{M} E\left[\frac{1}{\lambda(t_i;\theta)}\cdot\frac{\partial}{\partial\theta}\lambda(t_i;\theta)\right]. \tag{6.9}$$

To calculate the stochastic expectation appearing in the right-hand side of (6.9), the following corollary may be used [41].

Corollary 6.3. *Let $\{t_j\}_{j=1}^{N}$ be the photon TOAs with the pdf given in (3.10). Also, let*

$$g\left(\{t_j\}_{j=1}^{N}\right) \triangleq \prod_{j=1}^{N} r(t_j), \tag{6.10}$$

where $r(t)$ is any general function. Then, the stochastic expectation of $g\left(\{t_j\}_{j=1}^{N}\right)$ for a fixed N equals,

$$E_N\left[g\left(\{t_j\}_{j=1}^{N}\right)\right] = \frac{\mathrm{e}^{-\Lambda}}{N!}\Gamma^N, \tag{6.11}$$

where Λ is defined in (3.11), and,

$$\Gamma \triangleq \int_{t_0}^{t_f} r(t)\lambda(t)\mathrm{d}t. \tag{6.12}$$

Proof (Proof of Corollary 6.3). Let Ω_N be the event of receiving N photons at any N different increasing time instants in an interval $[t_0, t_f]$. Then, using the TOA pdf, given in (3.10),

$$E_N\left[g\left(\{t_j\}_{j=1}^{N}\right)\right] = \int_{\Omega_N} p\left(\{t_i\}_{i=1}^{N}, N\right)\prod_{j=1}^{N} r(t_j)\,\mathrm{d}t_1\cdots\mathrm{d}t_N, \tag{6.13}$$

where Ω_N is the event of receiving N photons at any N different increasing time instants in interval $[t_0, t_f]$. Note that the sequence $\{t_1, t_2, \dots, t_N\}$ is in increasing order. As there exists $N!$ permutations of t_i, for a fixed N, the probability that N number of t_i's occurs in no special order is $N!$ times that of the sequence occurring in increasing order. Therefore,

$$\begin{aligned}
&\int_{\Omega_N} p\left(\{t_i\}_{i=1}^N, N\right) \prod_{j=1}^N r(t_j)\,\mathrm{d}t_1 \cdots \mathrm{d}t_N \\
&= \frac{1}{N!} \int_{t_0}^{t_f} \int_{t_0}^{t_f} \cdots \int_{t_0}^{t_f} p\left(\{t_i\}_{i=1}^N, N\right) \prod_{j=1}^N r(t_j)\mathrm{d}t_1 \cdots \mathrm{d}t_N \\
&= \frac{\mathrm{e}^{-\Lambda}}{N!} \int_{t_0}^{t_f} \int_{t_0}^{t_f} \cdots \int_{t_0}^{t_f} \prod_{j=1}^N r(t_j)\lambda(t_j)\,\mathrm{d}t_1 \cdots \mathrm{d}t_N \\
&= \frac{\mathrm{e}^{-\Lambda}}{N!} \left(\int_{t_0}^{t_f} r(t)\lambda(t)\mathrm{d}t \right)^N \\
&\triangleq \frac{\mathrm{e}^{-\Lambda}}{N!} \Gamma^N.
\end{aligned} \tag{6.14}$$

□

Now it can be seen that the stochastic expectation in the right-hand side of (6.9) is a special case of (6.11) where $N = 1$. Hence,

$$\begin{aligned}
E\left[\frac{1}{\lambda(t_i;\theta)} \cdot \frac{\partial}{\partial\theta} \lambda(t_i;\theta) \right] &= \mathrm{e}^{-\Lambda}\Gamma \\
&= \mathrm{e}^{-\Lambda} \int_{t_0}^{t_f} r(t;\theta)\lambda(t;\theta)\mathrm{d}t,
\end{aligned} \tag{6.15}$$

where

$$r(t;\theta) = \frac{\dot{\lambda}(t;\theta)}{\lambda(t;\theta)} \tag{6.16}$$

which yields in,

$$E\left[\frac{1}{\lambda(t_i;\theta)} \cdot \frac{\partial}{\partial\theta} \lambda(t_i;\theta) \right] = \mathrm{e}^{-\Lambda} \left(\lambda(t_f;\theta) - \lambda(t_0;\theta) \right). \tag{6.17}$$

As Λ increases as a function of observation time, the term $\mathrm{e}^{-\Lambda}$ approaches zero, while $\lambda(t_f;\theta) - \lambda(t_0;\theta)$ remains finitely bounded in magnitude. Therefore, the right-hand side of (6.17) approaches zero. This means that the regularity condition (4.5) holds, and as a result, according to Theorem 6.2, the MLE $\hat{\theta}$ is asymptotically efficient. ■

Assuming the velocities of the vehicles are known, the Doppler frequencies do not need to be estimated, and the phase estimates may be obtained by solving the following optimization problem,

$$\hat{\phi}_k = \underset{\phi_k \in (0,1)}{\arg\max}\, \Psi(\phi_k), \quad k = 1, 2. \tag{6.18}$$

Furthermore, the pulse delay estimator is determined from (5.1).

The statistical properties of the resulting pulse delay estimator are presented by the following theorem [49].

Theorem 6.4. *The pulse delay estimator determined from (6.18) and (5.1) is* asymptotically efficient, *and its variance is given by (4.44).*

Proof. From Theorem 6.1, the phase estimators are asymptotically efficient. Hence, using (5.1), the pulse delay estimator is asymptotically efficient as well, and its variance equals $2\mathrm{CRLB}(\phi_0)/f_2^2$ where $\mathrm{CRLB}(\phi_0)$ is presented in (4.42). ■

6.3 Numerical Determination of the MLE

To find the MLE of a parameter θ, the likelihood function needs to be maximized. If the allowable values of θ lie in a certain interval then the maximum of likelihood function in that interval needs to be found. The safest way to do this is to perform a grid search over that interval. As long as the spacing between θ values is small enough, it is guaranteed to reach to the maximum of likelihood function. However, if the range of θ is not known or confined to a finite interval then a grid search may not be computationally feasible. In such situations, iterative optimization procedures are employed. A typical one is the Newton-Raphson method. In general, these methods result in the MLE if the initial guess is close to the true value. If not, they may not converge or only convergence to a local maximum is attained. An important point about MLE is that the likelihood function changes for each data set.

The iterative methods attempt to maximize the likelihood function by finding the zero of its derivative,

$$\frac{\partial}{\partial \theta} \ln p(x;\theta) = 0. \tag{6.19}$$

The Newton-Raphson iteration is [50],

$$\theta^{(k+1)} = \theta^{(k)} - \left(\frac{\partial^2}{\partial \theta \partial \theta^{\mathrm{T}}} \ln p(x;\theta) \right)^{-1} \frac{\partial}{\partial \theta} \ln p(x;\theta) \Bigg|_{\theta=\theta^{(k)}}, \tag{6.20}$$

where

$$\left[\frac{\partial^2}{\partial\theta\partial\theta^{\mathrm{T}}}\ln p(x;\theta)\right]_{ij} = \frac{\partial^2}{\partial\theta_i\partial\theta_j}\ln p(x;\theta) \tag{6.21}$$

is the Hessian of the likelihood function, and $\partial \ln p(x;\theta)/\partial\theta$ is the gradient vector. In implementing (6.20), inversion of the Hessian matrix is not required because it can be rewritten as,

$$\begin{aligned}&\frac{\partial^2}{\partial\theta\partial\theta^{\mathrm{T}}}\ln p(x;\theta)\Big|_{\theta=\theta^{(k)}}\theta^{(k+1)}\\&=\frac{\partial^2}{\partial\theta\partial\theta^{\mathrm{T}}}\ln p(x;\theta)\Big|_{\theta=\theta^{(k)}}\theta^{(k)}-\frac{\partial}{\partial\theta}\ln p(x;\theta)\Big|_{\theta=\theta^{(k)}}\end{aligned} \tag{6.22}$$

Regarding Newton-Raphson iterative procedure, one should notice that it may not converge. In particular, this may be the case if the Hessian matrix is small. Even if it converges, the point found may only be a local optimum. Hence, to avoid these possibilities, it is beneficial to use several starting points.

6.4 ML Computational Complexity Analysis

From (6.5), to construct the ML cost function for a fixed ϕ_0, M additions must be performed. Also, the cost function must be evaluated at the measured TOAs. If the rate function's analytical equation is known, this does not impose a significant amount of computational cost.

As the mathematical equation of the rate function is not usually available in practice, its value at the TOAs must be found by interpolation. If a simple linear interpolation is used, it costs four additions and two multiplications for each TOA. Hence, the total computational cost, for a fixed ϕ_0, is $6M$ flops. Similar to the NLS or CC cases, because the cost function is not concave in general, adopting a grid search optimization approach over $(0,1)$ cycle is desirable. If the search interval is divided into N_{g} grids, the total imposed computational cost due to the interpolation is $6MN_{\mathrm{g}}$ flops. This means that the amount of interpolation computations also grows as the number of received photons, M, increases.

6.5 Computational Complexity: Summary

Note that $M \gg N_p$ for long observations. Hence, compared to the NLS and CC cases, the amount of calculations for the ML estimator significantly increases as the observation time becomes longer. This is a noticeable disadvantage of the ML approach compared to the NLS estimation. This is to be avoided for implementation purposes. Hence, if the NLS or CC estimator's variances are within an acceptable

Table 6.1 Computational cost

Estimator	Additions	Subtractions	Multiplications	Divisions
CC	$N_b(N_p-1)+(N_b-1)N_g$	N/A	N_bN_g	$2N_b+1$
NLS	$N_b(N_p-1)+N_bN_g$	N_bN_g	N_bN_g	$2N_b$
ML	$M+4MN_g$	N/A	$2MN_g$	N/A

distance from the CRLB, it is more convenient to implement one of them. This is indeed the case. It is clear from (4.44) and (5.9) that the difference between the NLS estimator's variance and the CRLB becomes smaller as the observation increases. As a result, especially for long observation times, implementation of epoch-folding-based algorithms is computationally more efficient. Note that the ARE, given in (5.39), can also be calculated for each pulsar to measure how far is the estimator's performance from the CRLB.

If the cross-correlation function is calculated using the time domain approach, the computational cost for CC and NLS estimators is almost the same. As the correlation function can be found using the Fourier domain approach, its computational cost is expected to be less than the NLS estimator.

All computational cost results, presented in Sects. 5.6.2, 5.6.3, and 6.4, are summarized in Table 6.1.

6.6 Absolute Velocity Errors

Similar to the CC and NLS cases, if the velocity errors on each spacecraft are almost equal, $\Delta v_1 \approx \Delta v_2$, this results in the same deterioration of initial phase estimation on each detector. Hence, from (5.1), since the difference between the phase estimates is employed for estimation of the pulse delay, it remains asymptotically unbiased. However, the variance of estimation error becomes larger than the case where the absolute velocities are perfectly known. This error can be modeled by increasing the variance of the measurement noise in the Kalman filtering stage. If these assumptions are not valid, the absolute velocities can be estimated from (6.6).

6.7 Numerical Examples

The algorithm presented in Sect. 3.8 is used to generate the photon TOAs associated with the Crab pulsar (PSR B0531+21). The simulations are performed using the Monte-Carlo technique with over 500 independent realization of photon TOAs for each observation time. The constant arrival rates are chosen to be $\lambda_b = 5$ and $\lambda_s = 15$ ph/s.

First, the ML phase estimator is numerically simulated on one detector. Then the pulse delay performance is numerically examined. Finally, the computational cost needed for calculation of all of the proposed estimators is studied.

6.7.1 ML Phase Estimator

The spacecraft velocity on the direction vector pointing to the pulsar is assumed to be $v = 3$ km/s; and the initial phase observed at the detector is $\phi_1 = 0.2$ cycle.

Figure 6.1 shows the RMS error of the MLE. The ML estimates are determined by solving the optimization problem (6.18). A grid search approach on the interval $[0,1)$ cycle over 1,024 grids is utilized to solve (6.18). The ML plot shows that the estimator attains the CRLB for long observation times. Similar to the CC and NLS estimators, as the observation time drops below a certain threshold, the RMS error deviates from the CRLB. This behavior is due the fact that in this region, the maximum of the LLF does not lie in the vicinity of the true parameter and the estimator is biased.

The cost function (6.5) is plotted in Fig. 6.2 for different observation times. The plots show that as the observation time increases, the cost function becomes larger. Figure 6.2 also show that each of the cost functions has multiple maxima. Hence, they verify that, to avoid getting trapped in local maxima, adopting a grid search optimization approach is necessary. As the search is performed over the whole interval $(0,1)$ cycle, the algorithm initialization issues do not arise.

It is also verified that $\Lambda(\phi_0)$ in (6.3) can safely be dropped from the cost function. The ML phase estimates are found using (6.3) as the cost function. No meaningful difference is observed between the obtained RMS errors and the ones obtained using (6.5).

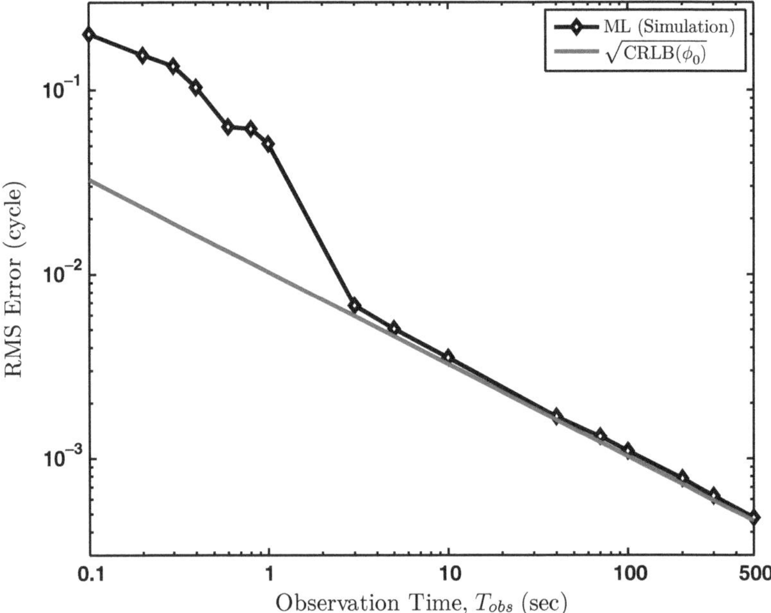

Fig. 6.1 RMS error for maximum-likelihood estimator (MLE)

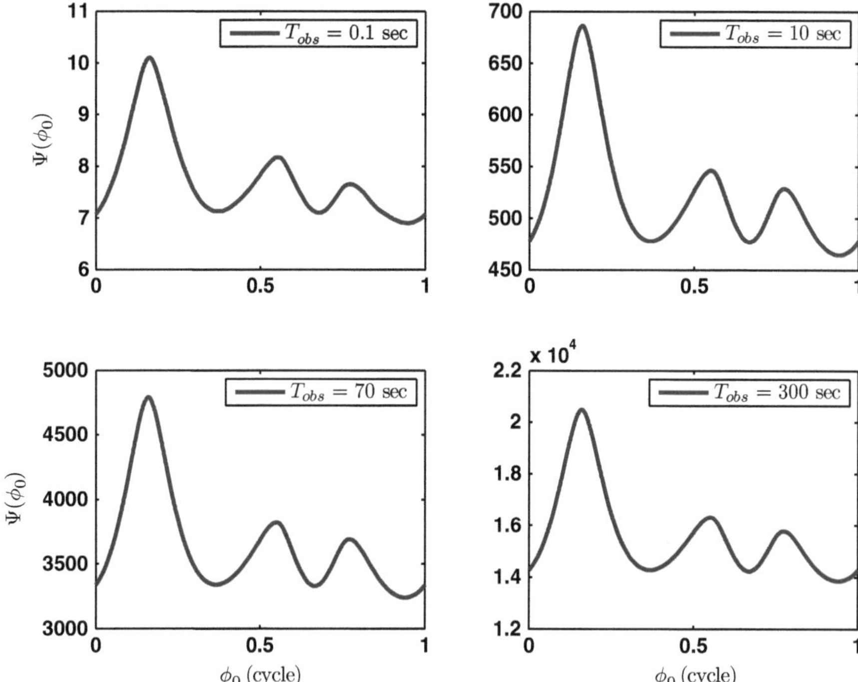

Fig. 6.2 The ML cost function

6.7.2 Pulse Delay Estimator

Assuming the velocity data is available, the pulse delay estimator is numerically evaluated. The spacecraft velocities are assumed to be $v_1 = 3$ km/s and $v_2 = 9$ km/s; and the initial phase observed at the first detector is $\phi_1 = 0.2$ cycle. The relative distance between the spacecraft is assumed to be 180 km. Hence, the pulse delay to be estimated is $t_d = 0.6$ ms.

Figure 6.3 shows the RMS of the proposed pulse delay estimator, obtained through simulation, against the CRLB, calculated as in (4.44). The plots show that as time goes on, the estimator asymptotically attains the CRLB. The cost functions (6.5) are optimized using a grid search approach in the domain of $(0,1)$ cycle. Similar to NLS and CC estimators, as the observation time is reduced, there is a threshold point at which the variance of the estimation error starts to deviate from the theoretical value. This is due to the fact that the estimator becomes biased if the observation time is not long enough. The reason is that as the observation time goes below this threshold, the number of detected photons is not high enough. Therefore, the realization of TOAs does not effectively represent the rate function of received photons. Hence, it results in a distorted cost function whose maximum does not lie in the vicinity of the true phase delay.

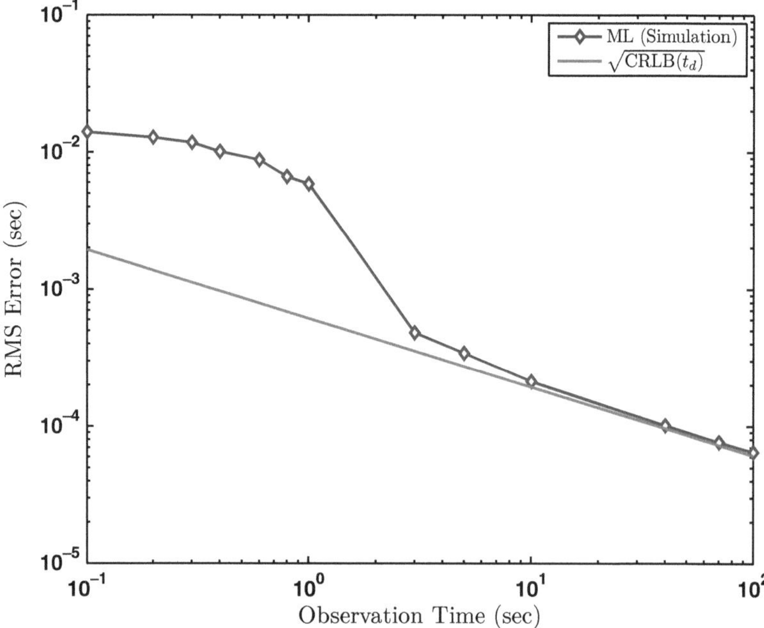

Fig. 6.3 RMS error for ML-based delay estimator

The performance of the proposed estimators is also studied when the absolute velocity data is not perfectly known. The errors are assumed to be $\Delta v_1 = 1{,}000$ and $\Delta v_2 = 1{,}200$ m/s. As Fig. 6.4 shows the RMS errors increase compared to the case where the velocity data is perfectly known.

6.7.3 Computational Cost

Figure 5.1 shows that asymptotic performance of the CC and NLS estimators is very close to the CRLB. This motivates the numerical implementation study of each estimator. Hence, the CPU time used by MATLAB to perform the calculations for one Monte Carlo realization is investigated. The utilized processor is an Intel 2.4 GHz dual core.

Figure 6.5 shows that the computational cost for linear interpolation used to calculate the CC estimator is almost independent of the observation time. This is to be expected, as the interpolation is performed at one point just to find a finer maximizer for the cross-correlation function. It also shows that the linear interpolation used for calculation of the MLE and the epoch folding have almost the same computational complexity, and as expected, the amount of calculations almost linearly grows as the observation time becomes longer.

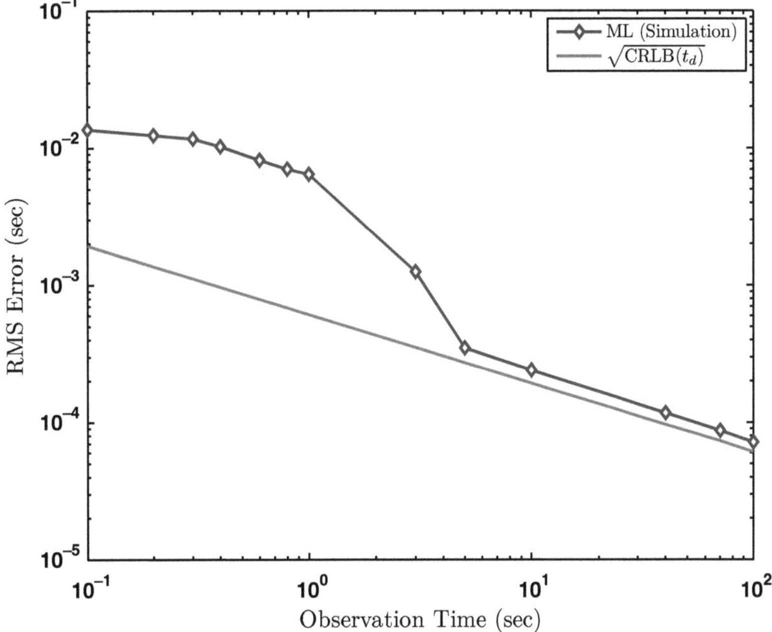

Fig. 6.4 RMS error for ML-based delay estimator in presence of spacecraft velocity errors

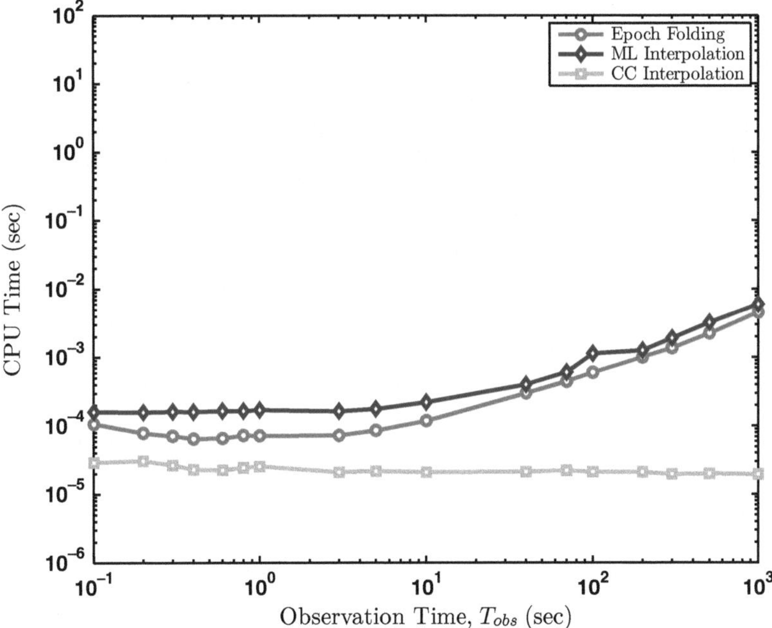

Fig. 6.5 CPU time used for epoch folding and interpolations

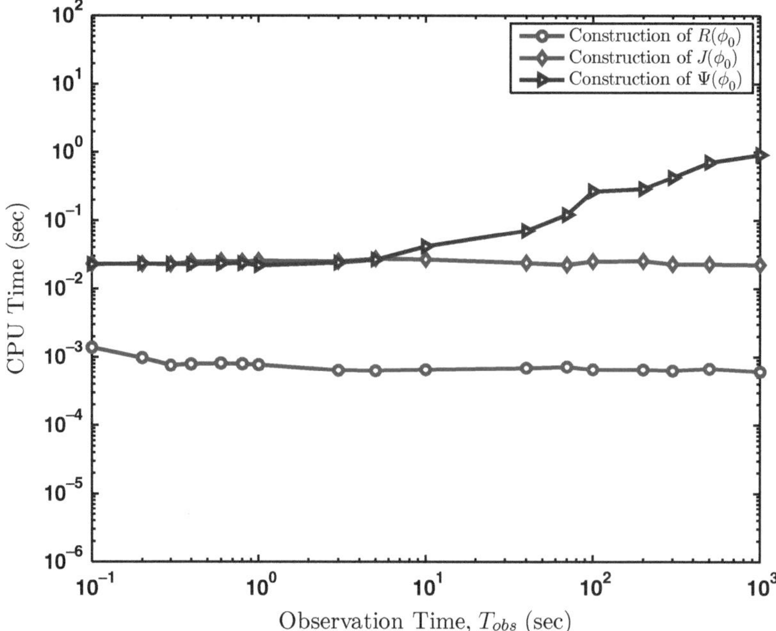

Fig. 6.6 CPU time used for construction of the cost functions

From Fig. 6.6, it can be seen that the amount of calculations to construct the CC and NLS cost functions does not change with the observation time. However, the computational cost to form the ML cost function almost linearly grows as the observation time becomes longer. It also shows that using Fourier domain approach to calculate the CC cost function significantly reduces the amount of calculations.

The total CPU time used to find the CC estimator is the sum of the epoch folding time, the time used for construction of the CC cost function, and the parabolic interpolation time. The total CPU time to find the NLS estimator is the sum of the epoch folding time and the cost function construction time for all grid points. The total CPU time for the MLE is the sum of the construction time and the total interpolation time for all grids. The reason is that for each grid, an interpolation is needed for the evaluation of the cost function at the measured TOAs. These plots are all shown in Fig. 6.7.

As the plots show, the ML CPU time is bigger than the one for the NLS algorithm, and it grows significantly faster with the observation time. The ML calculational cost rate with respect to the observation time is almost 15 times bigger than the NLS rate in this simulation scenario. Furthermore, the NLS CPU time is about 10 times bigger than the CPU time used to calculate the CC estimator.

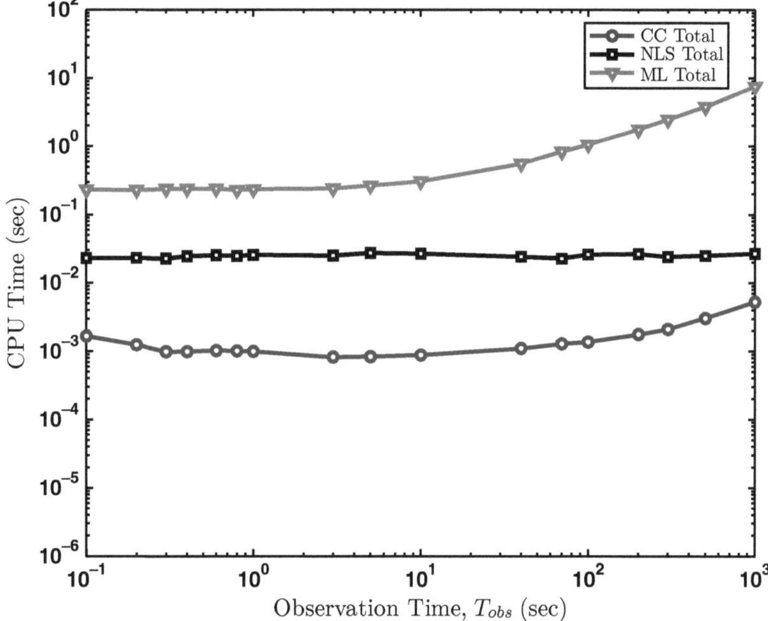

Fig. 6.7 Total CPU time used for calculation of the estimators

6.8 Summary

Using the pdf of the photon TOAs we construct an ML criterion, and formulate an asymptotically efficient estimator of the initial phase and the Doppler frequency for each detector. When the spacecraft velocity is known, it becomes only a phase estimator. Although the MLE asymptotically achieves the CRLB, we show that its computational cost is significantly higher than the epoch folding-based estimators. Under certain conditions, the pulse delay estimator can be used even when the spacecraft absolute velocities are not perfectly known. We also verify the analytical results on the MLE's performance and the computational costs via numerical examples.

Chapter 7
Recursive Estimation

7.1 Introduction

In this chapter, a recursive algorithm is formulated which can be used to find the relative navigation solution between the two spacecraft. The navigation system is equipped with IMUs which provide the spacecraft acceleration data. The dynamics of relative position between the two spacecraft and a model of the IMU accuracy are utilized for developing the navigation algorithm. The measurements, which are obtained by time tagging the photons, are modeled as a linear function of the projected relative position onto the unit direction pointing to the pulsar plus the measurement noise. The measurement noise variance is selected based on how well the pulse delay is estimated. Then, by applying a Kalman filter, the relative position, the relative velocity, the relative bias between accelerometers, and the differential time between clocks are estimated, and the steady state estimation error covariance is obtained. The effect of different system parameters on the achievable accuracy of relative position estimation is investigated. In particular, the effect of different values of IMU uncertainty, measurement noise variance, and number of pulsars used for estimation are considered.

We formulate the system dynamics for the navigation problem in Sect. 7.2. The measurement equation is presented in Sect. 7.3. The Kalman filter is given in Sect. 7.4. We investigate the observability of the system dynamics in Sect. 7.5. In Sect. 7.6, we briefly explain how the proposed approach is applicable to the absolute navigation problem. We offer a geometric approach for estimation of the spacecraft absolute velocities in Sect. 7.7. Section 7.8 provides different numerical simulation scenarios of the navigation system.

7.2 System Dynamics

X-ray pulsars are being observed over the sky map. All the measurements are performed in an inertial system whose origin is the SSB. The directional location of the pulsars is known. We denote this directional unit vector as $H_j^{(i)}$, where i identifies

A.A. Emadzadeh and J.L. Speyer, *Navigation in Space by X-ray Pulsars*,
DOI 10.1007/978-1-4419-8017-5_7,

the pulsar and j denotes the spacecraft. It is assumed that the space vehicles are close enough so that $H_1^{(i)} \cong H_2^{(i)} = H^{(i)}$.

Let $\Delta x \in \mathbb{R}^3$ be the relative position, $\Delta v \in \mathbb{R}^3$ be the relative velocity, and $\Delta a \in \mathbb{R}^3$ be the relative acceleration between the two spacecraft, and assume they follow the following dynamics

$$\Delta\dot{X} = A\Delta X + B\Delta a + Gw_v, \tag{7.1}$$

where $\Delta X = \left(\Delta x^{\mathrm{T}}\ \Delta v^{\mathrm{T}}\right)^{\mathrm{T}}$, the matrices are

$$A = \begin{pmatrix} 0_{3\times 3} & I \\ 0 & 0_{3\times 3} \end{pmatrix}, \quad B = \begin{pmatrix} 0_{3\times 3} \\ I_{3\times 3} \end{pmatrix}, \quad G = \begin{pmatrix} I_{3\times 3} \\ 0_{3\times 3} \end{pmatrix} \tag{7.2}$$

and w_v is an independent zero mean white noise process with a known power spectral density (PSD), denoted by W_v

$$E[w_v(t)w_v(\tau)^{\mathrm{T}}] = W_v\delta(t-\tau). \tag{7.3}$$

Each spacecraft is equipped with an inertial measurement unit which measures its acceleration, $a^{(i)}$, as

$$a_{\mathrm{IMU}}^{(i)} = a^{(i)} + w_a^{(i)} + b_a^{(i)}, \quad i = 1,2 \tag{7.4}$$

where $w^{(i)}$ is a white noise process with a known PSD, and $b_a^{(i)}$ is the accelerometer's bias. Let Δa_{IMU} be

$$\Delta a_{\mathrm{IMU}} \triangleq a_{\mathrm{IMU}}^{(1)} - a_{\mathrm{IMU}}^{(2)}. \tag{7.5}$$

Therefore, from (7.4)

$$\Delta a_{\mathrm{IMU}} = \Delta a + b_a + w_a, \tag{7.6}$$

where

$$\Delta a = a^{(1)} - a^{(2)} \tag{7.7}$$

and

$$b_a = b_a^{(1)} - b_a^{(2)} \tag{7.8}$$

is the relative bias. Because it slowly varies, and to keep the Kalman filter open, it is assumed to be a Brownian motion process, as

$$\dot{b}_a = w_b, \tag{7.9}$$

where w_a and w_b are zero mean independent white noise processes with known PSDs, denoted by W_{a} and W_{b}

$$E[w_a(t)w_a(\tau)^{\mathrm{T}}] = W_{\mathrm{a}}\delta(t-\tau) \tag{7.10}$$

$$E[w_b(t)w_b(\tau)^{\mathrm{T}}] = W_{\mathrm{b}}\delta(t-\tau). \tag{7.11}$$

The relative distance is determined by integration of Δa_{IMU},

$$\Delta \dot{X}_{\text{IMU}} = A\Delta X_{\text{IMU}} + B\Delta a_{\text{IMU}}, \tag{7.12}$$

where $\Delta X_{\text{IMU}} = \left(\Delta x_{\text{IMU}}^{\text{T}}\ \Delta v_{\text{IMU}}^{\text{T}}\right)^{\text{T}}$. Defining the error as,

$$X_{\text{e}} \triangleq \Delta X_{\text{IMU}} - \Delta X \tag{7.13}$$

from (7.1), (7.6), and (7.12), the dynamics of X_{e} are given by

$$\dot{X}_{\text{e}} = AX_{\text{e}} + Bb_a + Bw_a + Gw_v. \tag{7.14}$$

The differential time is common among all X-ray detectors. To model its slowly varying dynamics, and to keep the Kalman filter open (i.e., the Kalman gain does not converge to zero) [58], it is assumed to be a Brownian motion process,

$$\dot{t}_e = w_e, \tag{7.15}$$

where w_e is an independent zero mean white noise process with a known PSD, named W_e

$$E\left[w_e(t)w_e(\tau)^{\text{T}}\right] = W_e\delta(t-\tau). \tag{7.16}$$

Collecting the unknown vectors X_{e}, b_a, and t_e into the vector X as,

$$X \triangleq \begin{pmatrix} X_{\text{e}} \\ b_a \\ t_e \end{pmatrix} \tag{7.17}$$

its dynamics are given by

$$\dot{X} = FX + w, \tag{7.18}$$

where

$$F = \begin{pmatrix} A & B & 0 \\ 0_{3\times 6} & 0_{3\times 3} & 0 \\ 0_{1\times 6} & 0_{1\times 3} & 0 \end{pmatrix} \tag{7.19}$$

and

$$w \triangleq \begin{pmatrix} w_v \\ w_b \\ w_a \\ w_e \end{pmatrix} \tag{7.20}$$

is an independent white noise process where

$$E\left[w(t)w(\tau)^{\text{T}}\right] = \text{diag}(W_v, W_{\text{b}}, W_{\text{a}}, W_e)\delta(t-\tau). \tag{7.21}$$

To obtain the discrete model, the system (7.18) is sampled with the sampling time, T_s. The discrete model is as follows [59].

$$X(k+1) = \Phi X(k) + w_d(k) \tag{7.22}$$

where

$$\Phi = \exp(F T_\mathrm{s}) \tag{7.23}$$

and

$$w_d(k) = \int_0^{T_\mathrm{s}} \exp(F\tau) w(\tau)\, \mathrm{d}\tau. \tag{7.24}$$

Note that $F^3 = 0$. Therefore,

$$\Phi = I + T_\mathrm{s} F + \frac{T_\mathrm{s}^2}{2} F^2 \tag{7.25}$$

which equals

$$\Phi = \begin{pmatrix} I_{3\times3} & T_\mathrm{s} I & \frac{1}{2} T_\mathrm{s}^2 I & 0 \\ 0 & I_{3\times3} & T_\mathrm{s} I & 0 \\ 0 & 0 & I_{3\times3} & 0 \\ 0 & 0 & 0 & 1 \end{pmatrix}. \tag{7.26}$$

As $w(t)$ is zero mean, from (7.24), $w_d(k)$ is zero mean as well

$$E[w_d(k)] = 0 \tag{7.27}$$

and

$$E[\omega_d(k)\omega_d(j)^\mathrm{T}] = Q\delta_{kj}, \tag{7.28}$$

where Q, which is the variance matrix of the process noise, is given in (7.29).

$$Q = \begin{pmatrix} T_\mathrm{s} W_v + \frac{T_\mathrm{s}^3}{3} W_\mathrm{a} + \frac{T_\mathrm{s}^5}{20} W_\mathrm{b} & \frac{T_\mathrm{s}^2}{2} W_\mathrm{a} + \frac{T_\mathrm{s}^4}{8} W_\mathrm{b} & \frac{T_\mathrm{s}^3}{6} W_\mathrm{b} & 0 \\ \frac{T_\mathrm{s}^2}{2} W_\mathrm{a} + \frac{T_\mathrm{s}^4}{8} W_\mathrm{b} & T_\mathrm{s} W_\mathrm{a} + \frac{T_\mathrm{s}^3}{3} W_\mathrm{b} & \frac{T_\mathrm{s}^2}{2} W_\mathrm{b} & 0 \\ \frac{T_\mathrm{s}^3}{6} W_\mathrm{b} & \frac{T_\mathrm{s}^2}{2} W_\mathrm{b} & T_\mathrm{s} W_\mathrm{b} & 0 \\ 0 & 0 & 0 & T_\mathrm{s} W_e. \end{pmatrix} \tag{7.29}$$

7.3 Measurements

The pulse delay estimate for the ith pulsar is denoted by $\hat{t}_\mathrm{d}^{(i)}$. It is related to the relative position via

$$\begin{aligned} c\hat{t}_\mathrm{d}^{(i)}(k) &= c\hat{t}_x^{(i)} + ct_e(k) + \eta_d^{(i)}(k) \\ &= H^{(i)} \Delta x(k) + ct_e(k) + \eta_d^{(i)}(k), \end{aligned} \tag{7.30}$$

where $H^{(i)}$ is the directional unit vector pointing to the ith pulsar, and $\eta_d^{(i)}(k)$ is an independent zero-mean white noise sequence. The pulse delay estimates are available every T_m seconds. Hence, the autocorrelation function of $\eta_d^{(i)}(k)$ is

$$E[\eta_d^{(i)}(k)\,\eta_d^{(i)}(j)^{\mathrm{T}}] = R^{(i)}\delta_{kj},\ R^{(i)} = c^2\mathrm{var}[\hat{t}_{\mathrm{d}}]^{(i)}, \tag{7.31}$$

where $\mathrm{var}[\hat{t}_{\mathrm{d}}]^{(i)}$ is the variance of the ith pulse delay estimator, obtained every T_m seconds.

Let's define the measurement with respect to the ith pulsar as

$$y^{(i)}(k) \triangleq c\hat{t}_{\mathrm{d}}^{(i)}(k). \tag{7.32}$$

Using (7.30), the measurements have the following discrete model

$$y^{(i)}(k) = H^{(i)}\Delta x(k) + ct_e(k) + \eta_d^{(i)}(k). \tag{7.33}$$

Employing N different pulsars, N measurements are available, which can be collected into the following matrix format,

$$Y(k) = C\Delta X(k) + c\mathbb{1}t_e(k) + \eta_d(k), \tag{7.34}$$

where

$$Y \triangleq \begin{pmatrix} y^{(1)} \\ y^{(2)} \\ \vdots \\ y^{(N)} \end{pmatrix}, \quad C \triangleq \begin{pmatrix} H^{(1)} & 0_{1\times 3} \\ H^{(2)} & 0_{1\times 3} \\ \vdots & \\ H^{(N)} & 0_{1\times 3} \end{pmatrix} \tag{7.35}$$

and

$$\eta_d \triangleq \left(\eta_d^{(1)}\ \eta_d^{(2)}\ \ldots\ \eta_d^{(N)}\right)^{\mathrm{T}}. \tag{7.36}$$

The measurement noise, $\eta_d(k)$, is an independent white zero-mean process, and

$$E[\eta_d(k)\eta_d(j)^{\mathrm{T}}] = R\,\delta_{kj}, \tag{7.37}$$

where

$$R = \mathrm{diag}\left(R^{(i)}\right), \quad i = 1,\ldots,N \tag{7.38}$$

is the measurement noise variance matrix.

Recall that from (7.17), $\Delta X = \Delta X_{\mathrm{IMU}} - X_{\mathrm{e}}$. Therefore, (7.34) can be restated as

$$\begin{aligned} Z(k) &\triangleq -(Y(k) - C\Delta X_{\mathrm{IMU}}(k)) \\ &= CX_{\mathrm{e}}(k) - c\mathbb{1}t_e(k) + \eta(k), \end{aligned} \tag{7.39}$$

where $\eta(k) = -\eta_d(k)$, which is clearly an independent zero mean white noise with the same power spectral density as $\eta_d(k)$. Using (7.17), the measurement (7.39) can be expressed as a function of $X(k)$

$$Z(k) = HX(k) + \eta(k) \tag{7.40}$$

where,

$$H = \begin{pmatrix} C & 0_{N\times 3} & -c\mathbb{1} \end{pmatrix}. \tag{7.41}$$

To estimate the relative position, the relative velocity, the relative biases, and the differential time between clocks, the dynamic system in (7.22), and the measurement equation in (7.40) will be employed in a Kalman filter.

7.4 Discrete Time Estimation Process

Let $\hat{X}_k^{(-)}$ be the a priori system state estimate at time k, and $P_k^{(-)}$ be the a priori estimation error covariance, which are based on all measurements up to Z_{k-1}. Similarly, let $\hat{X}_k^{(+)}$ and $P_k^{(+)}$ denote the a posteriori estimate and covariance matrix after the measurement Z_k has been processed. Assuming that measurements start at $k=0$, the Kalman filter equations in discrete time are as follows [58, 59].

The filter's initial values are set as

$$\hat{X}_0^{(-)} = E[X_0] \tag{7.42a}$$

$$P_0^{(-)} = E[(X_0 - \hat{X}_0^{(-)})(X_0 - \hat{X}_0^{(-)})^{\mathrm{T}}] \tag{7.42b}$$

and the state estimates are updated as

$$K_k = P_k^{(-)} H^{\mathrm{T}} [H P_k^{(-)} H^{\mathrm{T}} + R]^{-1} \tag{7.43a}$$

$$\hat{X}_k^{(+)} = \hat{X}_k^{(-)} + K_k [Z_k - H\hat{X}_k^{(-)}] \tag{7.43b}$$

$$P_k^{(+)} = [I - K_k H] P_k^{(-)} [I - K_k H]^{\mathrm{T}} + K_k R K_k^{\mathrm{T}} \tag{7.43c}$$

$$\hat{X}_{k+1}^{(-)} = \Phi \hat{X}_k^{(+)} \tag{7.43d}$$

$$P_{k+1}^{(-)} = \Phi P_k^{(+)} \Phi^{\mathrm{T}} + Q. \tag{7.43e}$$

Note that the measurements (7.40) are available at a lower rate than the position estimate updates. Hence, when no measurement is available, the a priori and a posteriori estimates are equal. In other words, (7.43b) must be replaced by

$$\hat{X}_k^{(+)} = \hat{X}_k^{(-)} \tag{7.44}$$

in the meantime between the pulsar measurements (7.40) become available.

7.5 Discussion

For the Kalman filter to estimate all the unknown states, the dynamic system (7.22) and (7.40) must be observable. Hence, the observability matrix is investigated. It equals

$$\begin{aligned}\mathscr{O} &= \begin{pmatrix} H \\ HF \\ HF^2 \end{pmatrix} \\ &= \begin{pmatrix} \Gamma & 0 & 0 & -c\mathbb{1} \\ 0 & \Gamma & 0 & 0 \\ 0 & 0 & \Gamma & 0 \end{pmatrix} \end{aligned} \tag{7.45}$$

where

$$\Gamma_{N\times 3} = \begin{pmatrix} H^{(1)} \\ H^{(2)} \\ \vdots \\ H^{(N)} \end{pmatrix}. \tag{7.46}$$

Note that $\mathscr{O}$ is an $3N \times 10$ matrix. If there are at least four different pulsar measurements ($N \geq 4$) where no two are along the same directional vector from the spacecraft, then $\mathscr{O}$ will be full column-rank (i.e., rank 10). Hence the continuous system (7.18) is observable using the measurement (7.40). To conclude about the observability of discrete dynamics (7.22) and (7.40), the following theorem is employed [60].

Theorem 7.1. *Suppose the continuous system (7.18) and (7.40) is observable. A necessary and sufficient for its discretized equations (7.22) and (7.40), with sampling period T_s, to be observable is that $|\mathfrak{Im}[\lambda_i - \lambda_j]| \neq 2\pi m/T_s$ for $m = 1, 2, \ldots$, whenever $\mathfrak{Re}[\lambda_i - \lambda_j] = 0$, where λ_i is an eigenvalue of F.*

Because all eigenvalues of the matrix F are zero, then

$$\begin{aligned}|\mathfrak{Im}[\lambda_i - \lambda_j]| &= 0 \\ &\neq \frac{2\pi m}{T_s}. \end{aligned} \tag{7.47}$$

Therefore, it can be concluded that the discretized system is fully observable too, if $\mathscr{O}$ is full column-rank.

As a result, by choosing at least four different pulsars with different direction vectors, all the states can be estimated using the Kalman filter.

7.6 Absolute Navigation

The proposed navigation approach is applicable for absolute navigation as well. It is clear that if the location of one of the spacecraft is known, the relative navigation problem boils down to an absolute navigation problem. As the location of the SSB is known, placing one of the spacecraft at the SSB, using the same problem formulation, the absolute navigation solution can be obtained.

7.7 A Geometric Approach for Estimation of Absolute Velocities

In this section, we propose a different formulation of the navigation problem, using geometric dispersion of the X-ray pulsars over the sky map.

For simplicity, a two-dimensional scenario where four pulsars are utilized is considered (see Fig. 7.1). The ith pulsar's estimated time delay is composed of the true pulse delay and the differential time between clocks. From (5.1), and using Taylor series, it equals

$$\begin{aligned}\hat{t}_{\mathrm{d}}^{(i)} &= t_x^{(i)} + t_e \\ &= \frac{\Delta\hat{\phi}^{(i)} + n^{(i)}}{(1 + V_j \cos\theta_i / c) f_{\mathrm{s}}^{(i)}} \\ &\approx \frac{\Delta\hat{\phi}^{(i)} + n^{(i)}}{f_{\mathrm{s}}^{(i)}} \left(1 - \frac{V_j \cos\theta_i}{c}\right),\end{aligned} \tag{7.48}$$

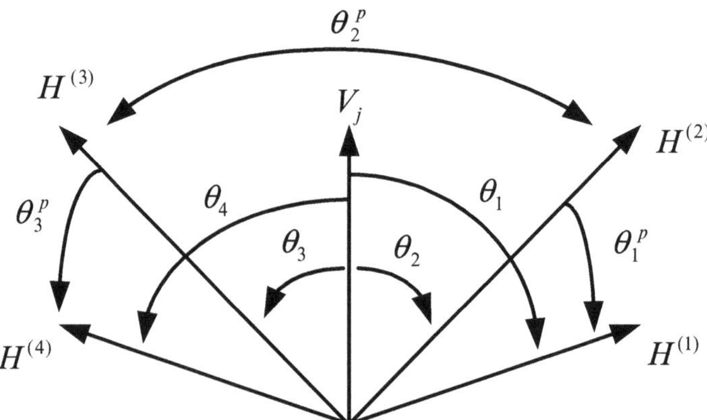

Fig. 7.1 Using pulsar geometry to estimate absolute velocities

where $\Delta\hat{\phi}^{(i)} = \hat{\phi}_1^{(i)} - \hat{\phi}_2^{(i)}$, V_j is the absolute velocity vector of the jth spacecraft, θ_i is the angle between V_j and $H^{(i)}$, and $n^{(i)}$ is the uncertainty modeling for the phase estimation error. Hence,

$$\frac{\Delta\hat{\phi}^{(i)}}{f_s^{(i)}} = t_x^{(i)} + t_e + \frac{\Delta\hat{\phi}^{(i)}}{f_s^{(i)} c} V_j \cos\theta_i + \eta^{(i)}, \tag{7.49}$$

where

$$\eta^{(i)} = -n^{(i)} \left(1 - \frac{V_j \cos\theta_i}{c}\right). \tag{7.50}$$

Defining the ith pulsar measurement as $z^{(i)} = \Delta\hat{\phi}^{(i)}/f_s^{(i)}$, (7.49) becomes

$$z^{(i)} = \hat{t}_x^{(i)} + t_e + z^{(i)} \frac{V_j \cos\theta_i}{c} + \eta^{(i)}. \tag{7.51}$$

Because $\hat{t}_x^{(i)} = H^{(i)}\Delta x/c$, the new set of measurements is

$$z^{(i)} = \frac{H^{(i)}\Delta x}{c} + t_e + z^{(i)} \frac{V_j \cos\theta_i}{c} + \eta^{(i)}. \tag{7.52}$$

Furthermore, the angular relations between directional vectors form a set of pseudo-measurements

$$\theta_1^{\mathrm{p}} = \theta_1 - \theta_2 + \gamma_1 \tag{7.53a}$$
$$\theta_2^{\mathrm{p}} = \theta_2 + \theta_3 + \gamma_2 \tag{7.53b}$$
$$\theta_3^{\mathrm{p}} = \theta_4 - \theta_3 + \gamma_3, \tag{7.53c}$$

where γ_i is the uncertainty. Using (7.52) and (7.53) as measurements, a new Kalman filter can be formulated for estimation of Δx, V_j, and θ_i.

7.8 Numerical Examples

To simulate the three-dimensional navigation algorithm, eight pulsars are selected (see Fig. 7.2 for pulsar geometry). The pulsar profiles are shown in Fig. 4.2. The pulsar galactic coordinates, and their photon flux values are given in Table 7.1 [19].

The detector area is assumed to be 10^4 cm^2. The effective pulsar rate, λ_s, is calculated according to the pulsar photon flux and the detector area. The effective background rate is chosen as $\lambda_b = \lceil \lambda_s/10 \rceil$. The corresponding values are given in Table 7.2.

In three different scenarios, the measurements are incorporated every 100 s, and every 500 s. The NLS-based pulse delay estimation approach is utilized, and

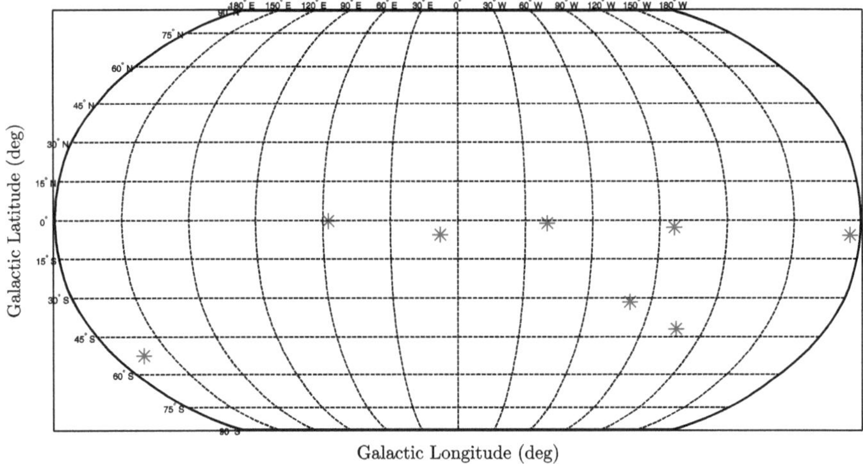

Fig. 7.2 Pulsar geometry on the sky map

Table 7.1 Employed pulsars

Pulsar	Period (s)	Galactic longitude (○)	Galactic latitude (○)	Flux (2–10 keV) (ph/cm^2/s)
B0531 + 21	0.0335	184.56	−5.78	1.54E + 00
B0540 − 69	0.0504	279.72	−31.52	5.15E − 03
B0833 − 45	0.0893	263.55	−2.79	1.59E − 03
B1509 − 58	0.1502	320.32	−1.16	1.62E − 02
B1821 − 24	0.0031	7.80	−5.58	1.93E − 04
B1937 + 21	0.0016	57.51	−0.29	4.99E − 05
B1055 − 52	0.1971	164.50	−52.45	1.64E − 06
J0437 − 47	0.0057	253.39	−41.96	6.65E − 05

Table 7.2 Employed pulsars (detector area: 10^4 cm^2)

Pulsar	λ_b (ph/s)	λ_s (ph/s)	σ_m (m) ($T_{obs} = 100$ s)	σ_m (m) ($T_{obs} = 500$ s)
B0531 + 21	1.54E + 03	1.54E + 04	5.64E + 2	2.52E + 2
B0540 − 69	5.15E + 00	5.15E + 01	3.46E + 4	1.54E + 4
B0833 − 45	1.59E + 00	1.59E + 01	4.67E + 4	2.09E + 4
B1509 − 58	1.62E + 01	1.62E + 02	8.62E + 4	3.86E + 4
B1821 − 24	1E + 00	1.93E + 00	4.56E + 3	2.04E + 3
B1937 + 21	1E + 00	4.99E − 01	4.53E + 3	2.02E + 3
B1055 − 52	1E + 00	1.64E − 02	1.67E + 8	7.49E + 7
J0437 − 47	1E + 00	6.65E − 03	1.11E + 5	4.99E + 4

$\sigma_m^{(i)} = \sqrt{R^{(i)}}$ values corresponding to each scenario are given in Table 7.2. The spacecraft absolute velocity data are not perfectly known. The velocity errors are assumed to be $\Delta v_1 = 1000$ and $\Delta v_2 = 1200$ m/s. Note that because of the effect of absolute velocity errors on the quality of measurements, the measurement noise variance values are bigger than their corresponding analytical values when the velocity data is perfectly known.

To initiate the Kalman filter, the *a priori* covariance matrix is chosen as

$$P_0 = \text{diag}(P_{0x}, P_{0v}, P_{0b_a}, P_{0t_e}), \tag{7.54}$$

where

$$\sqrt{P_{0x}} = \text{diag}(10^6, 10^6, 10^6) \quad (\text{m}) \tag{7.55a}$$
$$\sqrt{P_{0v}} = \text{diag}(10^4, 10^3, 10^2) \quad (\text{m/s}) \tag{7.55b}$$
$$\sqrt{P_{0b_a}} = \text{diag}(10^{0.5}, 10^{0.5}, 10^{0.5}) \quad (\text{m/s}^2) \tag{7.55c}$$
$$\sqrt{P_{0t_e}} = 10^3 \quad (\text{s}). \tag{7.55d}$$

The a priori state estimate at $k = 0$ is $\hat{X}_0^{(-)} = 0$. Note that although the initial uncertainties given in (7.55) change the transient response of the Kalman filter, they do not affect its steady state performance.

The IMU process noise PSDs used by the Kalman filter are given in (7.56). Note that to simulate the dynamics (7.22), it is assumed that $W_\text{b} = 0$ and $W_e = 0$, since they are assumed to be constant states. Although, they follow slowly time varying dynamics in practice. Hence, to take this fact into account, and to keep the Kalman filter open, they are modeled as Brownian motion processes in the estimation stage.

$$\sqrt{W_v} = 10^{-6} I \quad (\text{m}/\sqrt{\text{s}}) \tag{7.56a}$$
$$\sqrt{W_\text{a}} = 10^{-7} I \quad (\text{m}/\sqrt{\text{s}^3}) \tag{7.56b}$$
$$\sqrt{W_\text{b}} = 10^{-5} I \quad (\text{m}/\sqrt{\text{s}^5}) \tag{7.56c}$$
$$\sqrt{W_e} = 10^{-6} \quad (\sqrt{\text{s}}) \tag{7.56d}$$

The units in (7.56) are obtained from (7.29), noting that the diagonal elements have units m^2, $(\text{m/s})^2$, $(\text{m/s}^2)^2$, and s^2, respectively.

In (7.56), the accelerometer bias uncertainty, W_b, is chosen small based on the fact that the available IMUs are very accurate. These values are easily achievable by available commercial IMUs such as LN200 [61]. Choosing other W_b values only affects achievable estimation accuracies. The accelerometer uncertainty, W_a, represents the applied forces to the spacecraft from the celestial sources (ex. solar radiation pressure). Hence, it is chosen small as well.

To find out how accurate the relative distance between the spacecraft can be estimated, the state estimation error, and the standard deviation (STD) of the a posteriori estimation error are investigated. In other words, the square roots of the diagonal elements of $P_k^{(+)}$ are considered. Note that using (7.17), the standard deviation of $\hat{X}_e$ equals to the standard deviation of $\Delta\hat{X}$, which represents the relative position and relative velocity vectors between the spacecraft. All results are obtained through Monte Carlo simulation over 3,000 independent realization of the stochastic signals. Different sampling times, T_s, are chosen for discretizing the continuous dynamics. As expected, T_s does not change either the error covariance in the steady state or the time that the steady state is reached; however, the number of iterations to get to the same values of the error covariance varies. The initial state of the dynamic system (7.22) is assumed to be a Gaussian random variable

$$X_0 \sim \mathcal{N}(0, P_0), \tag{7.57}$$

where P_0 is given in (7.54).

In all scenarios the Monte Carlo state estimation error, obtained by averaging over 3,000 realizations, is observed to be zero. Hence, its plots are omitted. The Monte Carlo a posteriori standard deviation, along with the analytical one obtained from (7.43c) are plotted, and it is shown that they satisfactorily match. The analytical values that are computed in the Kalman filter are based on modeling some of the variables as Brownian motion, as opposed to the simulation model where the variables are assumed to be slowly varying (in some cases constants). As a result, the simulation STD plots may not perfectly match the analytical values, but they are very close.

The estimation error obtained from one [example] realization is plotted as well. As expected, we can see that at any time epoch, most of the Mote Carlo estimation error points are within the 1-σ envelope.

In the first scenario, eight pulsars listed in Table 7.1 are selected, and the measurements are obtained every 100 s. The results are plotted in Figs. 7.3–7.6. The obtained estimation accuracies are given in Tables 7.3, 7.4, and 7.5.

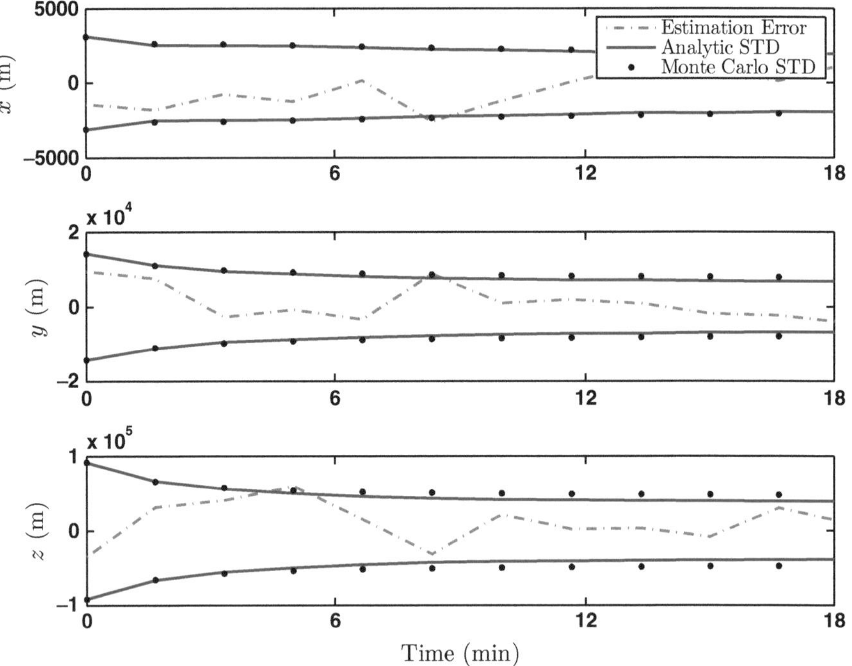

Fig. 7.3 Relative position estimation ($N = 8$, $T_{\text{obs}} = 100$ s)

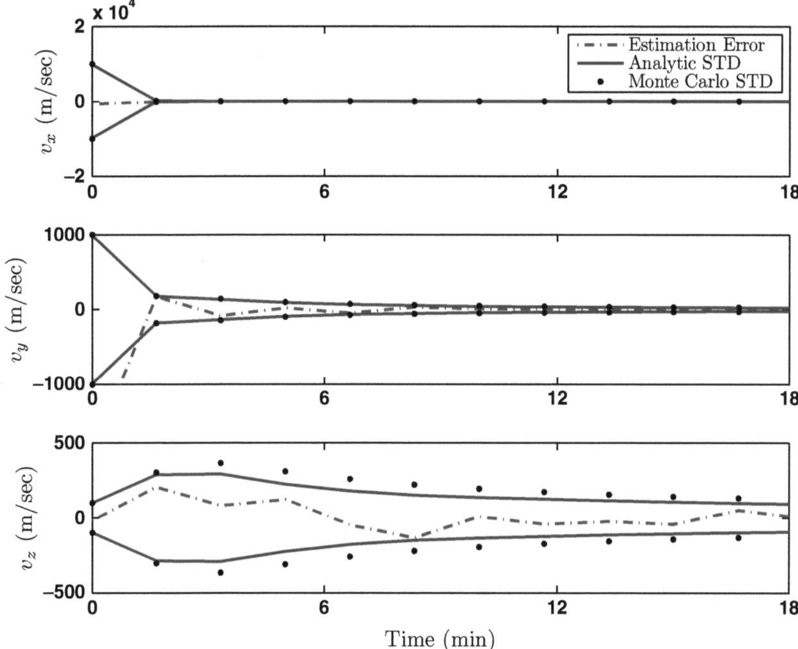

Fig. 7.4 Relative velocity estimation ($N = 8$, $T_{\text{obs}} = 100\,\text{s}$)

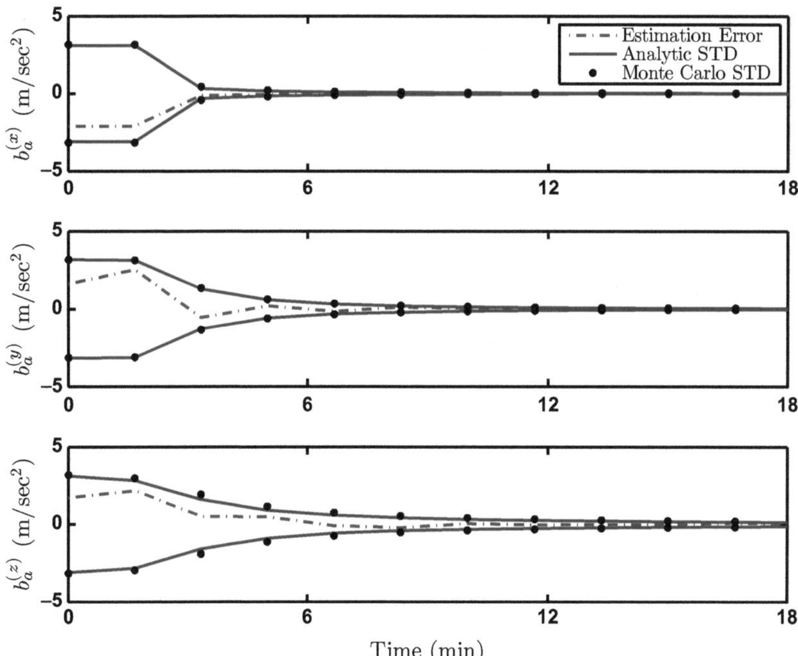

Fig. 7.5 Relative bias estimation ($N = 8$, $T_{\text{obs}} = 100\,\text{s}$)

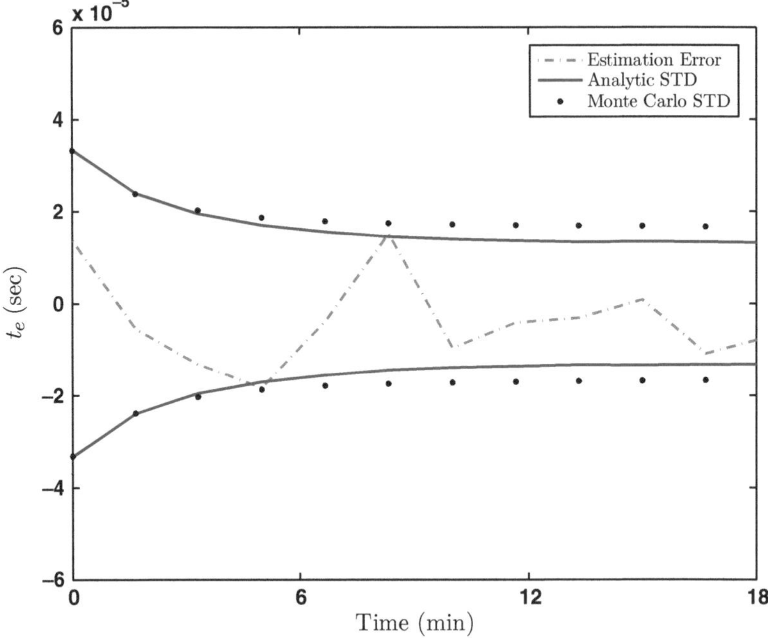

Fig. 7.6 Clock differential time estimation ($N = 8$, $T_{\text{obs}} = 100$ s)

Table 7.3 Position estimation results

Scenario	$\sqrt{P_x}$ (m)	$\sqrt{P_y}$ (m)	$\sqrt{P_z}$ (m)
$N = 8$, $T_{\text{obs}} = 100$ s	1.93E+3	6.74E+3	3.87E+4
$N = 8$, $T_{\text{obs}} = 500$ s	1.16E+3	4.97E+3	3.21E+4
$N = 4$, $T_{\text{obs}} = 100$ s	2.16E+4	4.12E+4	6.83E+4

Table 7.4 Velocity estimation results

Scenario	$\sqrt{P_{v_x}}$ (m/s)	$\sqrt{P_{v_y}}$ (m/s)	$\sqrt{P_{v_z}}$ (m/s)
$N = 8$, $T_{\text{obs}} = 100$ s	7.52	21.87	92.96
$N = 8$, $T_{\text{obs}} = 500$ s	2.09	7.91	47.55
$N = 4$, $T_{\text{obs}} = 100$ s	15.21	114.78	183.08

Table 7.5 IMU Bias and clock differential time estimation results

Scenario	$\sqrt{P_b^{(x)}}$ (m/s^2)	$\sqrt{P_b^{(y)}}$ (m/s^2)	$\sqrt{P_b^{(z)}}$ (m/s^2)	$\sqrt{P_e}$ (sec)
$N = 8$, $T_{\text{obs}} = 100$ s	1.31E−2	3.74E−2	1.34E−1	1.32E−5
$N = 8$, $T_{\text{obs}} = 500$ s	1.59E−3	5.87E−3	3.45E−2	1.14E−5
$N = 4$, $T_{\text{obs}} = 100$ s	2.05E−2	1.80E−1	2.15E−1	8.01E−5

In the second scenario, employing the same pulsars, the measurements are provided every 500 s. Hence, the measurement noise STD values decrease. The corresponding values are given in Table 7.2. As Figs. 7.7–7.10 show, using more accurate measurements, results in obtaining more accurate estimates. The estimation results are given in Tables 7.3, 7.4, and 7.5.

In the third scenario, to investigate the effect of the pulsar geometry on the estimation accuracy, just the first four pulsars from Table 7.1 are chosen for navigation. The measurements are updated every 100 s. The plots are shown in Figs. 7.11–7.14, and the resulted accuracies are given in Table 7.3. As expected, compared to the case where eight pulsars were employed, the estimation error standard deviations have increased (Tables 7.3, 7.4, and 7.5).

Recall from (7.33) that the relative position, Δx, and the differential time between clocks, t_e, show up directly in the measurements. Hence, we expect the Kalman filter to be able to estimate them immediately, and it takes longer for the relative bias and velocity estimates to converge to the steady state values. From the a priori initial standard deviations, given in (7.55), and using the estimation plots shown in Figs. 7.3–7.14, this can be verified where the a posteriori position and differential time estimation standard deviations dramatically drop when the first measurements is incorporated at $t = 0$.

Another interesting point regarding Tables 7.3, 7.4, and 7.5 is that the x-coordinate estimates are more accurate than the y-coordinate estimates. Considering Fig. 7.2, we expect this phenomenon. As the sky map shows, the employed pulsars are geometrically more distributed along the x-axis than the y-axis.

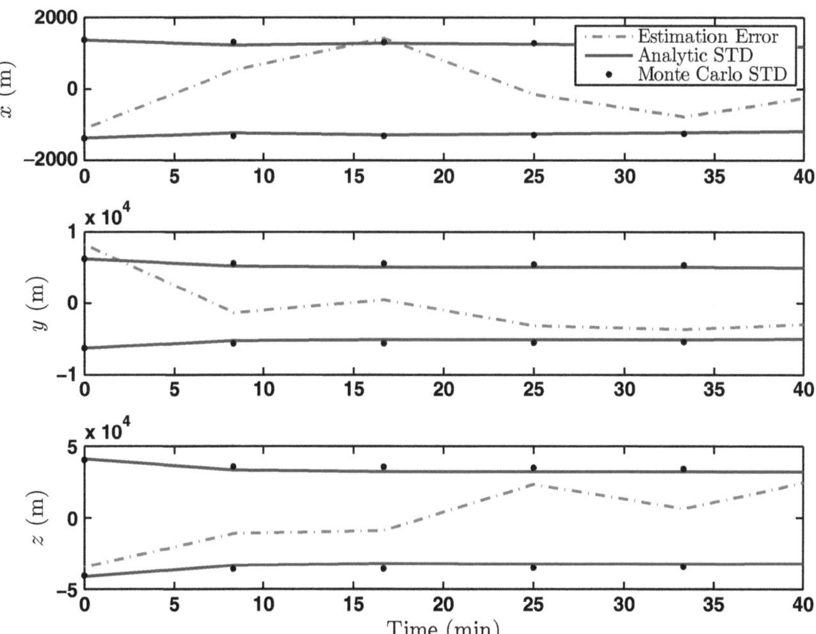

Fig. 7.7 Relative position estimation ($N = 8$, $T_{\text{obs}} = 500$ s)

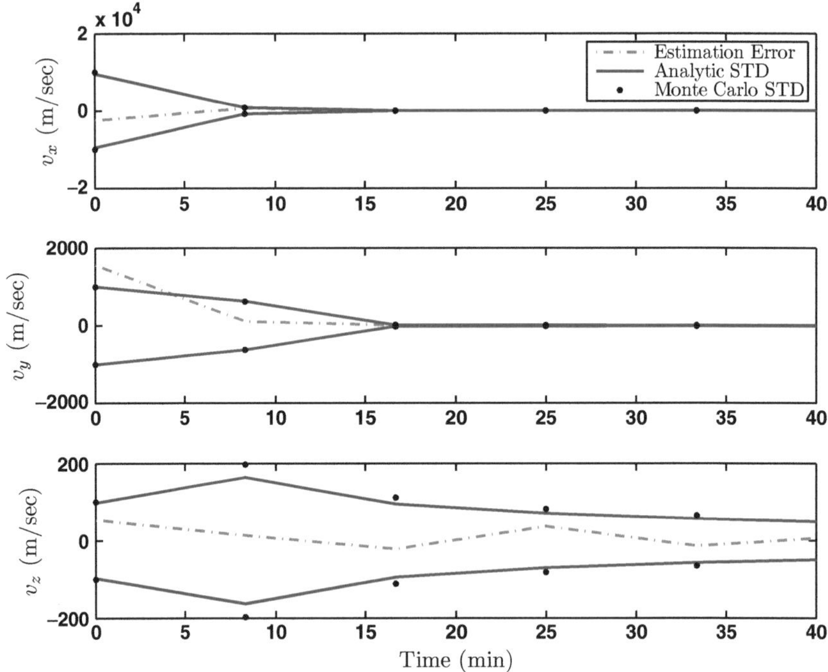

Fig. 7.8 Relative velocity estimation ($N = 8$, $T_{\text{obs}} = 500$ s)

Furthermore, Tables 7.3, 7.4, and 7.5 show that the z-coordinate estimates are less accurate than the x- and y-coordinate estimates. This implies that the pulsars geometric distribution along the z-axis is worse than the other axes.

In summary, we investigated effectiveness of the proposed algorithm through different simulation scenarios. Achievable estimation accuracies depend on different elements of the navigation system. The main factors are the pulse delay estimation accuracies, the IMU accuracies, and the pulsars geometric distribution on the sky map.

We have also performed a parametric study to investigate the range of achievable accuracies as a function of the quality of measurements. All simulation parameters are the same as the scenario where $N = 8$ and $T_{\text{obs}} = 500$ s. The only difference is the steady state STD of estimation errors are obtained for $k_R\sigma_m$ in the range of $0.1 \leq k_R \leq 10$, where σ_m values are the measurement noise standard deviations for $T_{\text{obs}} = 500$ s, given in Table 7.2. The results are plotted in Figs. 7.15–7.18. From these plots, the range of estimation accuracies are given in Table 7.6.

These ranges show that employing bright pulsars, position estimation accuracies in the order of a couple of 100 m are achievable for position estimation. The velocity estimation accuracies in the order of a few meters per second can be obtained. The attainable bias estimation accuracy is less than 1 mm/s^2. Furthermore, the differential time between clocks can be estimated in the order of a few micro seconds.

The effect of the IMU uncertainty on relative navigation solution is also investigated. All parameters are the same as the scenario where $N = 8$ and $T_{\text{obs}} = 500$ s.

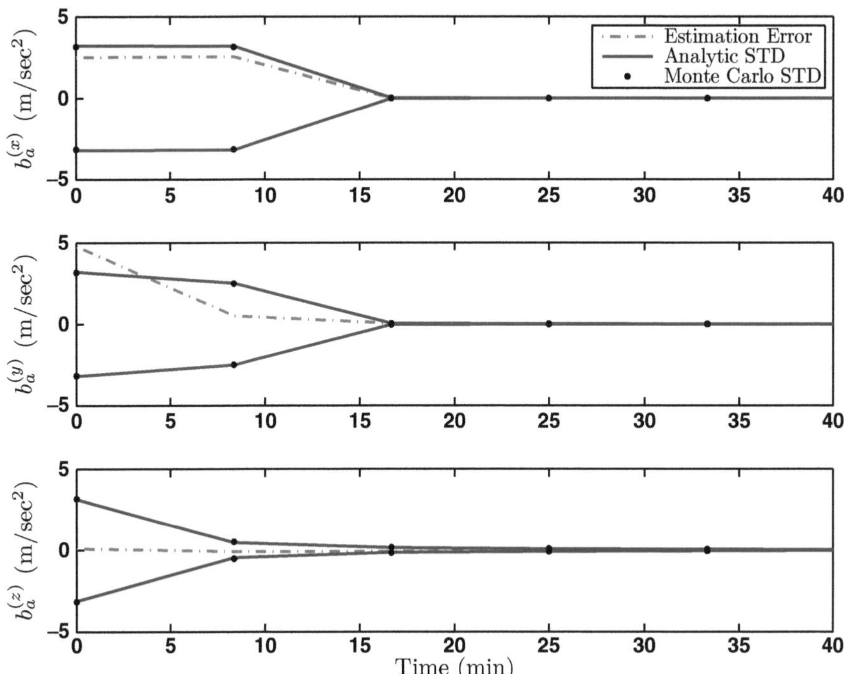

Fig. 7.9 Relative bias estimation ($N = 8$, $T_{\text{obs}} = 500\,\text{s}$)

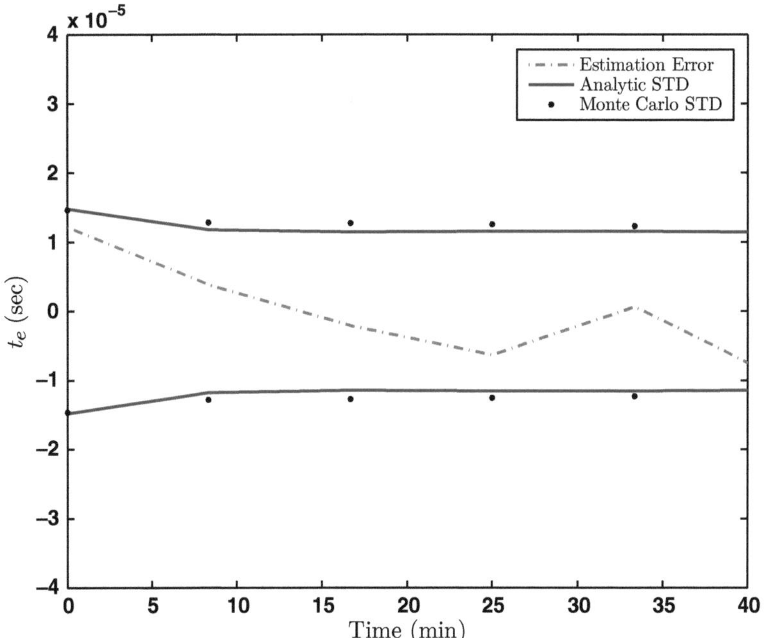

Fig. 7.10 Clock differential time estimation ($N = 8$, $T_{\text{obs}} = 500\,\text{s}$)

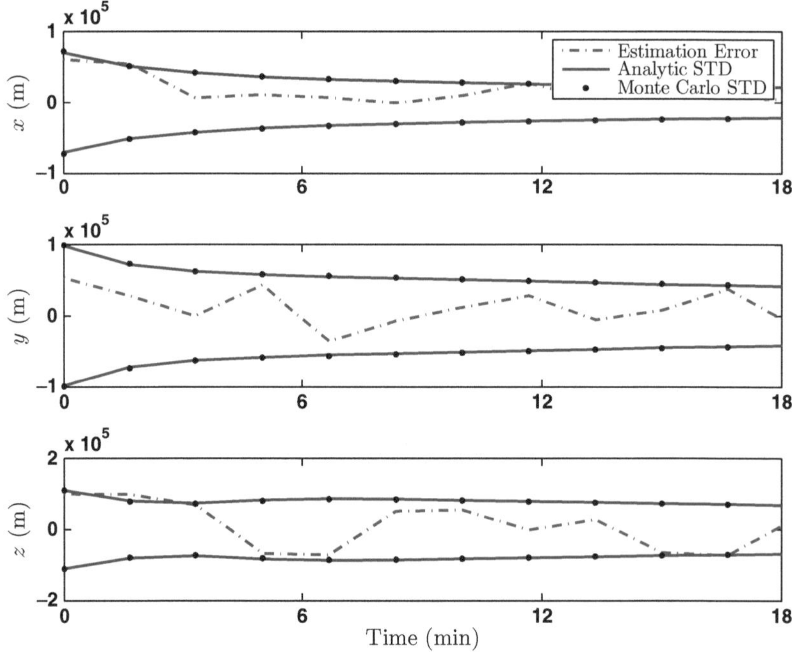

Fig. 7.11 Relative position estimation ($N = 4$, $T_{\text{obs}} = 100\,\text{s}$)

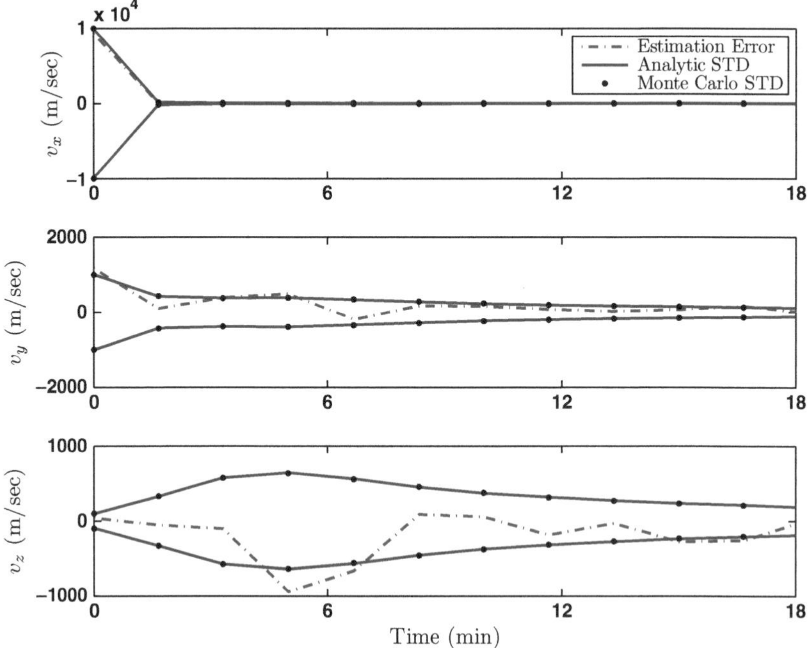

Fig. 7.12 Relative velocity estimation ($N = 4$, $T_{\text{obs}} = 100\,\text{s}$)

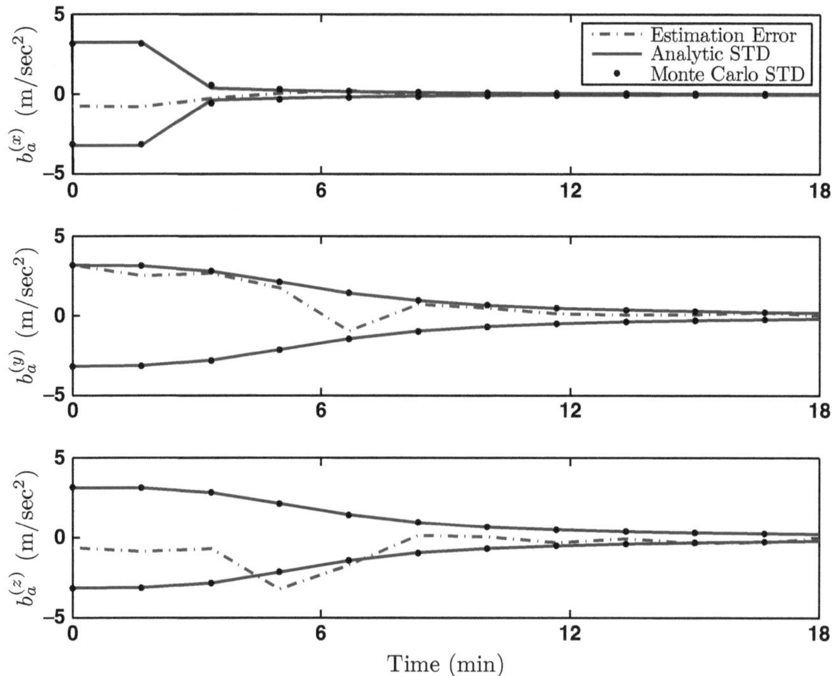

Fig. 7.13 Relative bias estimation ($N = 4$, $T_{\text{obs}} = 100\,\text{s}$)

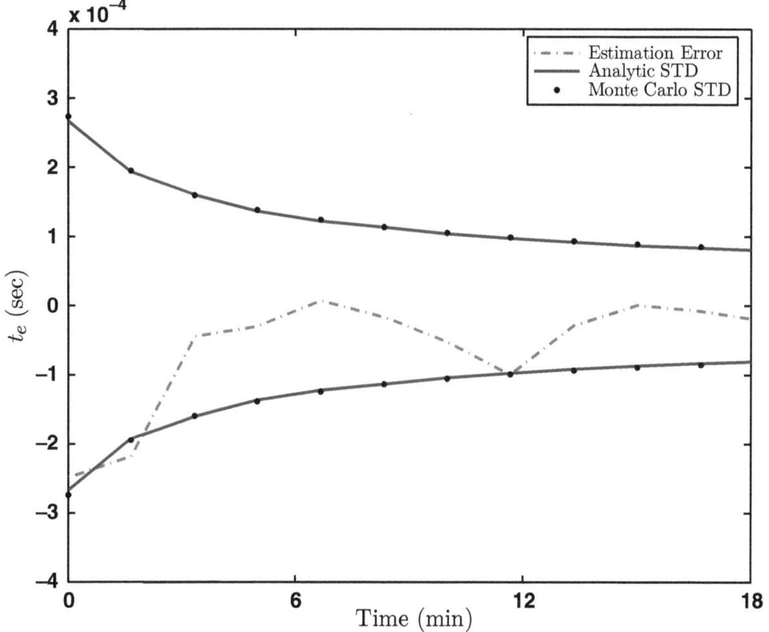

Fig. 7.14 Clock differential time estimation ($N = 4$, $T_{\text{obs}} = 100\,\text{s}$)

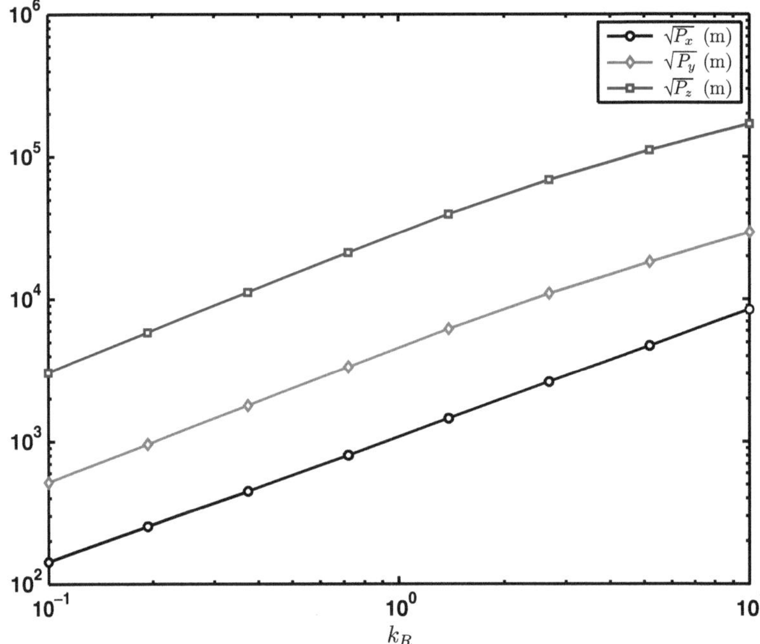

Fig. 7.15 The STD of position estimation error

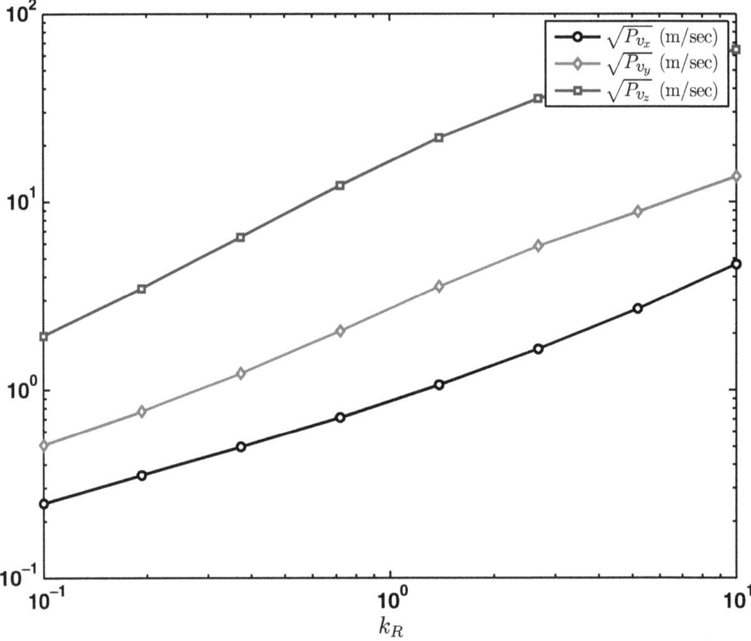

Fig. 7.16 STD of velocity estimation error

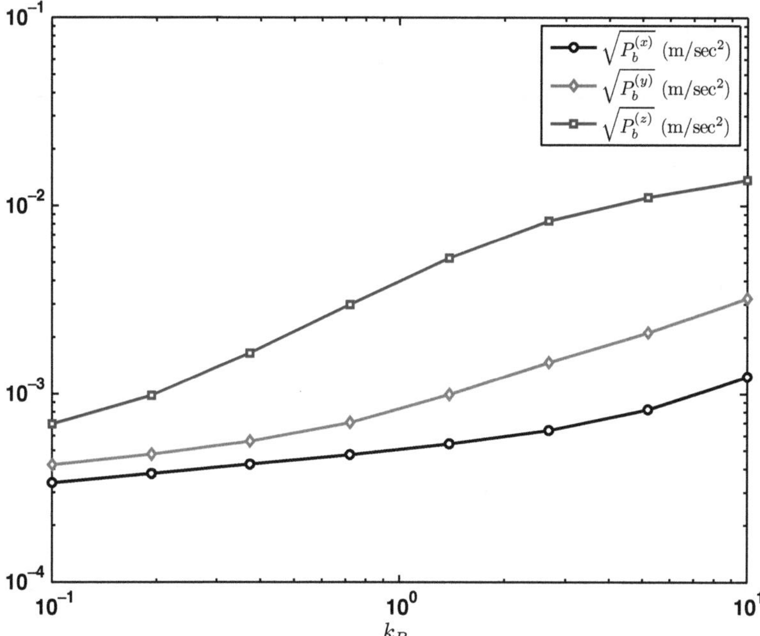

Fig. 7.17 The STD of bias estimation error

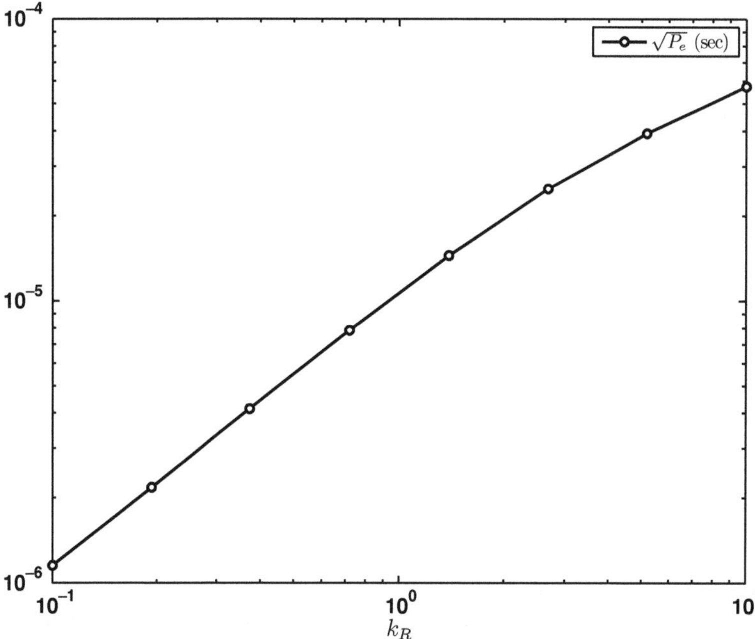

Fig. 7.18 The STD of differential time estimation error

Table 7.6 Estimation results when R or Q matrices change

	R varies		Q varies	
	Min	Max	Min	Max
$\sqrt{P_b^{(x)}}$ (m)	142.42	8438.4	852.27	1423.6
$\sqrt{P_b^{(y)}}$ (m)	516.83	2946.8	2947.5	5157.0
$\sqrt{P_b^{(z)}}$ (m)	3051.2	169950	17137.0	30406.0
$\sqrt{P_{v_x}}$ (m/s)	0.2489	4.6571	0.4737	2.4889
$\sqrt{P_{v_y}}$ (m/s)	0.5076	13.6063	1.3696	5.0678
$\sqrt{P_{v_z}}$ (m/s)	1.9286	64.1779	7.0152	19.1681
$\sqrt{P_b^{(x)}}$ (m/s^2)	3.3694E−4	0.0012	0.12515	0.0034
$\sqrt{P_b^{(y)}}$ (m/s^2)	4.1967E−4	0.0032	0.3271	0.0042
$\sqrt{P_b^{(z)}}$ (m/s^2)	6.9230E−4	0.0137	0.0016	0.0069
$\sqrt{P_e}$ (s)	1.1508E−6	5.7419E−5	5.7494E−6	1.1473E−5

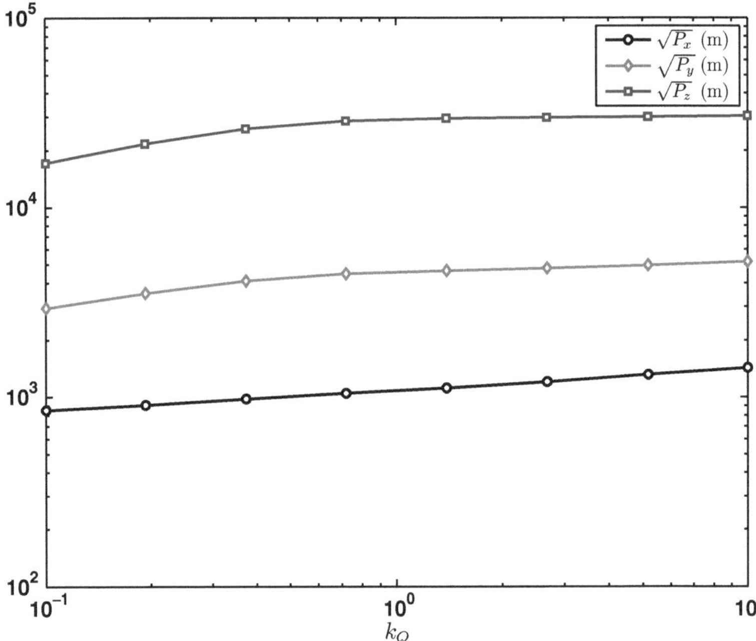

Fig. 7.19 The STD of position estimation error

The only parameters that are changed are the IMU uncertainties. Let σ_p be any of the IMU uncertainties given in (7.56d). Then, the achievable estimation accuracies are examined for $k_Q\sigma_p$, where $0.1 \le k_Q \le 10$. The results are plotted in Figs. 7.19–7.22. The minimum and maximum accuracies are also given in Table 7.6.

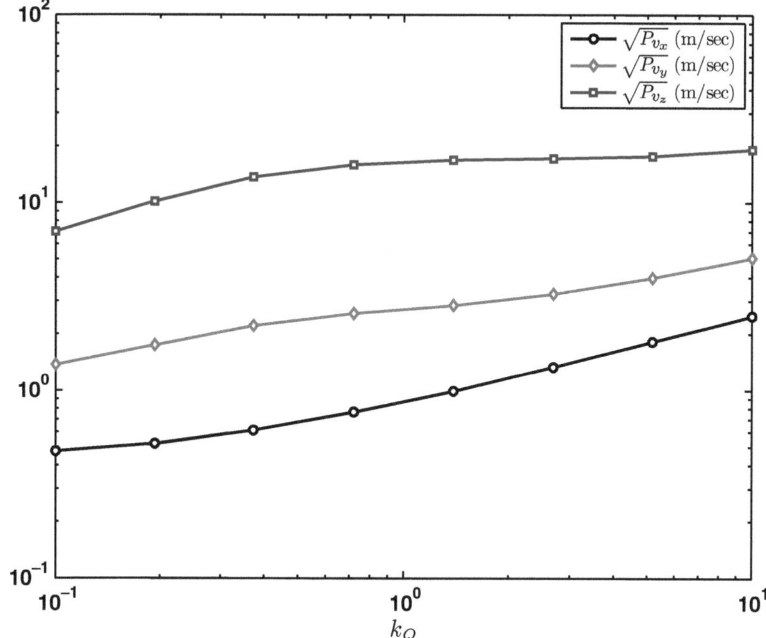

Fig. 7.20 The STD of velocity estimation error

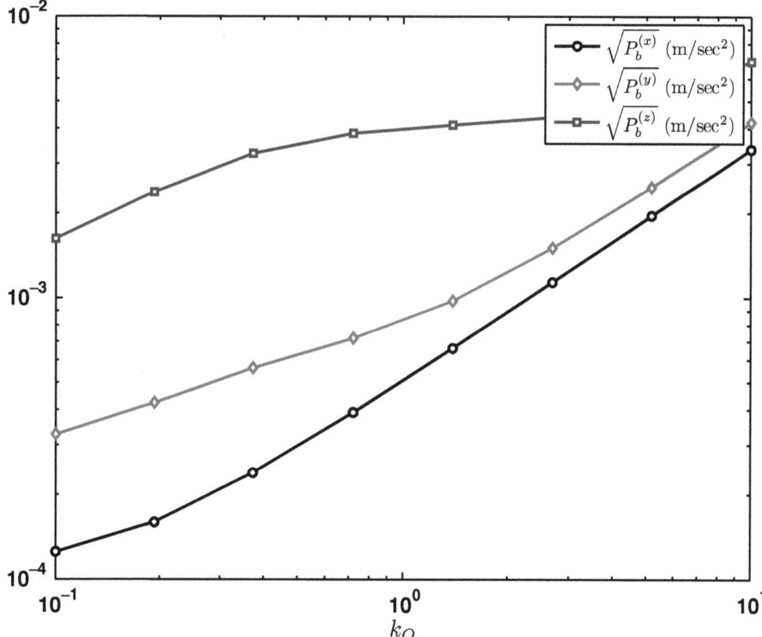

Fig. 7.21 The STD of bias estimation error

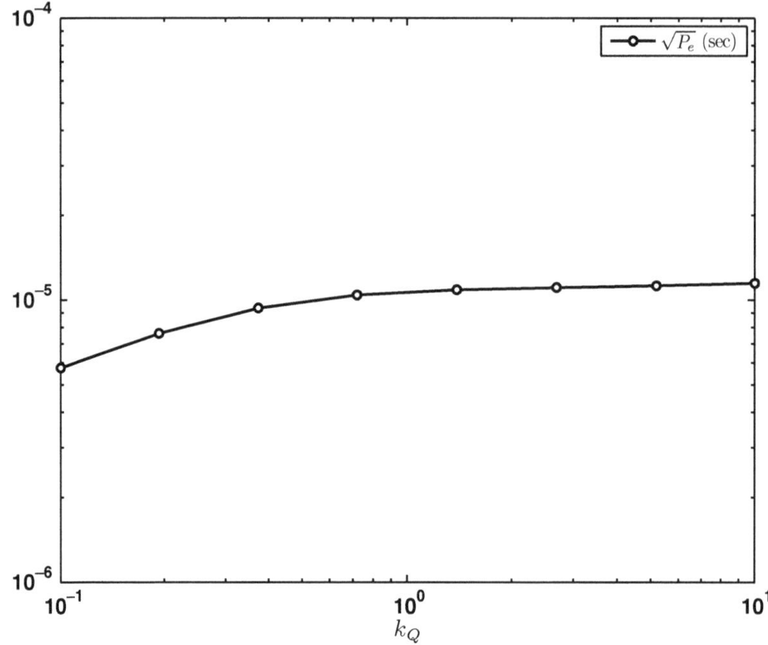

Fig. 7.22 The STD of differential time estimation error

7.9 Summary

In this chapter, we present a recursive navigation algorithm. We utilize the pulse delay estimates to construct the measurements for a Kalman filter. We suggest to employ IMUs on each spacecraft to provide the acceleration data. Using the Kalman filter, we offer a recursive algorithm to estimate the relative position, the relative velocity, the relative accelerometer biases, and the differential time between clocks. Through different simulation scenarios, applicability of the proposed navigation algorithm is also verified.

Chapter 8
Epilog

This book has proposed a new approach for navigation of spacecraft in space employing X-ray pulsar measurements. The presented approach is applicable for both absolute and relative navigation problems. Pulsars emitting in the X-ray band were chosen because of their stable period and their geometric distribution in the sky map. The main advantage of using X-ray pulsars for navigation is that relatively small size detectors can be used for detection of the X-ray photons on board a spacecraft. This facilitates the spacecraft design procedure.

The developed navigation technique is based on utilizing X-ray detectors on each spacecraft and locking the detectors on the same pulsar. Hence, the vehicle farther from the pulsar detects a signal whose intensity is the time delayed version of the one detected by the closer vehicle. The distance between the space vehicles is proportional to the time delay. The proposed approach is to periodically estimate the pulse delay, and then use these estimates as measurements in a recursive algorithm to find the navigation solution. To analyze the system, mathematical models were developed for the X-ray pulsar signals, and the Cramér–Rao lower bound for estimation of the pulse delay was presented. Two different strategies for estimation of the pulse delay were suggested. One strategy was to employ epoch folding. The procedure, which is used to recover photon intensities, was mathematically studied and characterized. Two different estimators based on epoch folding were formulated and their asymptotic performance was analyzed. One estimator uses the cross-correlation function between the empirical rate function and the known pulsar intensity function. The other estimator is obtained through solving a nonlinear least squares problem. It was shown that these estimators are consistent, but not asymptotically efficient. The second strategy was to directly utilize the measured photon time of arrivals. Using this strategy and based on a maximum likelihood criterion, a pulse delay estimator was formulated, and it was shown that the estimator is asymptotically efficient. It was shown that the cross-correlation-based estimator is computationally more efficient than the other two estimators.

Space vehicles are equipped with inertial measurement units to provide the acceleration information, which are converted to velocity data and utilized by the pulse delay estimator. The pulse delay estimates, in turn, are taken in as measurements by the Kalman filter for recursive estimation. The measurement noise variance is selected based on accuracy of the pulse delay estimates. Models of spacecraft

A.A. Emadzadeh and J.L. Speyer, *Navigation in Space by X-ray Pulsars*,
DOI 10.1007/978-1-4419-8017-5_8,

dynamics and inertial measurement units are employed by the proposed algorithm. It was shown that the relative inertial measurement unit biases and the differential time between detectors' clocks can be estimated as well as the relative position and velocity between the spacecraft. The three-dimensional relative navigation solution can be obtained by taking measurements on four or more different X-ray pulsars. Depending on the number of pulsars, their characteristics, their geometric distribution in the sky, and inertial measurement unit uncertainties, the achievable estimation accuracies were examined.

To enhance the position estimation accuracy, improving models of the spacecraft motion and IMU dynamics is necessary for future work. Because some models are now nonlinear, employing filtering techniques such as the extended Kalman filter and the unscented Kalman filter may be appropriate. Modeling clock errors and studying their effect on the navigation solution are necessary to obtain more accurate results. Furthermore, to estimate the system biases with more precision, smoothing algorithms may be utilized. Developing signal processing techniques for estimation of the pulse phase when spacecraft velocities are varying rapidly is another necessity for improving the navigation algorithm. Additionally, it is of interest to investigate the navigation problem in more detail in situations where Doppler frequencies and pulse phases are simultaneously estimated. Proposing pulse delay estimators for these situations and analysis of their performance are important areas of future research. Addressing the cycle ambiguity problem is another topic to work on. Also, taking advantage of pulsar geometry for new formulations of the navigation problem, is an interesting field for enhancing the pulsar-based navigation solution.

References

1. J. F. Jordan, "Navigation of Spacecraft on Deep Space Missions," *Journal of Navigation*, vol. 40, no. 1, pp. 19–29, Jan. 1987.
2. W. G. Melbourne, "Navigation Between the Planets," *Scientific American*, vol. 234, no. 6, pp. 58–74, Mar. 1976.
3. J. R. Wertz, *Spacecraft Attitude Determination and Control*. Dordrecht: Kluwer, 1978.
4. R. R. Bate, D. D. Muller, and J. E. White, *Fundamentals of Astrodynamics*. New York: Dover, 1971.
5. R. H. Battin, *An Introduction to the Mathematics and Methods of Astrodynamics*. Washington D.C.: American Institute of Aeronautics and Astronautics, Revised Edition, 1999.
6. B. W. Parkinson and J. J. Spiker Jr., *Global Positioning System*. Washington D.C.: American Institute of Aeronautics and Astronautics, Inc., 1996, vols. I and II.
7. About The Deep Space Network (2010). Jet Propulsion Lab (JPL), California Institute of Technology, Pasadena, CA. [Online]. Available: http://deepspace.jpl.nasa.gov/dsn/
8. A. A. Emadzadeh, C. G. Lopes, and J. L. Speyer, "Online Time Delay Estimation of Pulsar Signals for Relative Navigation using Adaptive Filters," in *IEEE/ION PLANS*, Monterey, CA, USA, May 2008, pp. 714–719.
9. A. A. Emadzadeh and J. L. Speyer, "A Study of Pulsar Signal Modeling and its Time Delay Estimation for Relative Navigation," in *ION National Technical Meeting (NTM)*, San Diego, CA, USA, 2008, pp. 119 - 126.
10. D. Folta, C. Gramling, A. Long, D. Leung, and S. Belur, "Autonomous Navigation Using Celestial Objects," in *Americal Astronautical Society (AAS)*, Girdwood, Alaska, USA, Aug. 1999, pp. 2161–2177.
11. R. Gounley, R. White, and E. Gai, "Autonomous Satellite Navigation by Stellar Refraction," *Journal of Guidance, Control and Dynamics*, vol. 7, no. 2, pp. 129–134, 1984.
12. D. Lorimer (2001). Binary and Millisecond Pulsars at the New Millenium. [Online]. Available: http://relativity.livingreviews.org/Articles/lrr-2001-5/
13. ATNF Pulsar Catalogue (2010). [Online]. Available: http://www.atnf.csiro.au/research/pulsar/psrcat/
14. A. G. Lyne and F. Graham-Smith, *Pulsar Astronomy*, 3rd ed. Cambridge: Cambridge University Press, 2006.
15. D. N. Matsakis, J. H. Taylor, T. M. Eubanks, and T. Marshall, "A Statistic for Describing Pulsar and Clock Stabilities," *Astronomy and Astrophysics*, vol. 326, pp. 924–928, Oct. 1997.
16. J. G. Hartnett and A. Luiten (2010). "A Comparison of Astrophysical and Terrestrial Frequency Standards: Which are the best clocks?". [Online]. Available: http://arxiv.org/abs/1004.0115v2
17. P. A. Charles and F. D. Seward, *Exploring the X-ray Universe*. Cambridge: Cambridge University Press, 1995.
18. W. Becker and J. Trumper, "The X-ray Luminosity of Rotation-Powered Neutron Stars," *Astronomy and Astrophysics*, vol. 326, pp. 682–691, 1997.

A.A. Emadzadeh and J.L. Speyer, *Navigation in Space by X-ray Pulsars*,
DOI 10.1007/978-1-4419-8017-5,

19. S. I. Sheikh, "The Use of Variable Celestial X-ray Sources for Spacecraft Navigation," Ph.D. dissertation, University of Maryland, 2005.
20. V. Pulsar: Chandra Reveals a Compact Nebula Created by a Shooting Neutron Star (2010). [Online]. Available: http://chandra.harvard.edu/photo/2000/vela/
21. Nasa's Fermi Telescope Probes Dozens of Pulsars (2010). [Online]. Available: http://www.nasa.gov/mission_pages/GLAST/news/pulsar_passel.html
22. NRAO, Green Bank Telescopes (2010). [Online]. Available: http://www.gb.nrao.edu
23. NAIC, arecibo Observatory Home (2010). [Online]. Available: http://www.naic.edu
24. M. D. Griffin and J. R. French, *Space Vehicle Design*. Washington D.C.: American Institute of Aeronautics and Astronautics, 1991.
25. P. Ray, K. Wood, and B. Phlips, "Spacecraft Navigation Using X-ray Pulsars," *Naval Research Lab (NRL) Review*, pp. 95–102, 2006.
26. P. Reichley, G. Downs, and G. Morris, "Use of Pulsar Signals as Clocks," *NASA Jet Propulsion Laboratory Quarterly Technical Review*, vol. 1, no. 1, pp. 80–86, July 1971.
27. G. Downs and P. E. Reichley, "Techniques for Measuring Arrival Times of Pulsar Signals I: DSN Observations from 1968 to 1980," *NASA Jet Propulsion Laboratory, California Institute of Technology, Pasadena CA, NASA Technical Reports NASA-CR-163564*, Aug. 1980.
28. D. W. Allan, "Millisecond Pulsar Rivals Best Atomic Clock Stability," in *41st Annual Symposium on Frequency Control. 1987*, 1987, pp. 2–11.
29. G. S. Downs, "Interplanetary Navigation Using Pulsating Radio Sources," in *NASA Technical Reports N74-34150*, Oct. 1974, pp. 1–12.
30. T. J. Chester and S. A. Butman, "Navigation Using X-ray Pulsers," in *NASA Technical Reports N81-27129*, Oct. 1981, pp. 22–25.
31. K. Wallace, "Radio Stars, What they are and The Prospects for their Use in Navigational Systems," *Journal of Navigation*, vol. 41, no. 3, pp. 358–374, 1988.
32. J. E. Hanson, "Principles of X-ray Navigation," Ph.D. dissertation, Stanford University, 1996.
33. J. Sala, A. Urruela, X. Villares, R. Estalella, and J. M. Paredes (2004). "Feasibility Study for a Spacecraft Navigation System relying on Pulsar Timing Information." [Online]. Available: http://www.esa.int/gsp/ACT/phy/pp/pulsar-navigation.htm
34. S. I. Sheikh, A. R. Golshan, and D. J. Pines, "Absolute and Relative Position Determination Using Variable Celestial X-ray Sources," in *30th Annual AAS Guidance & Control Conference*, Breckenridge, Colorado, USA, Feb. 2007, pp. 855–874.
35. S. I. Sheikh, P. S. Ray, K. Weiner, M. T. Wolff, and K. S. Wood, "Relative Navigation of Spacecraft Utilizing Bright, Aperiodic Celestial Sources," *63rd Annual Meeting of Institute of Navigation (ION)*, pp. 444–453, April 2007.
36. S. I. Sheikh, D. J. Pines, S. R. Ray, K. Wood, M. N. Lovellette, and M. T. Wolff, "Spacecraft Navigation Using X-ray Pulsars," *Journal of Guidance, Control and Dynamics*, vol. 29, no. 1, pp. 49–63, 2006.
37. S. I. Sheikh and D. J. Pines, "Recursive Estimation of Spacectraft Position and Velocity Using X-ray Pulsar Time of Arrival Measurements," *NAVIGATION: Journal of The Institute of navigation*, vol. 53, no. 3, pp. 149–166, 2006.
38. D. W. Woodfork, "The Use of X-ray Pulsars for Aiding GPS Satellite Orbit Determination," Master's thesis, Air Force Institue Of Technology, Wright-Patterson Air Force Base, Ohio, 2005.
39. A. A. Emadzadeh, "Relative Navigation between Two Spacecraft Using X-ray Pulsars," Ph.D. dissertation, University of California, Los Angeles, 2009.
40. S. M. Ross, *Introduction to Probability Models*, 6th ed. New York: Academic Press, 1996.
41. I. Bar-David, "Communication Under the Poisson Regime," *IEEE Transactions on Information Theory*, vol. 15, no. 1, pp. 31–37, Jan. 1969.
42. A. R. Golshan and S. I. Sheikh, "On Pulsar Phase Estimation and Tracking of Variable Celestial X-ray Sources," in *63rd Annual Meeting of Institute of Navigation (ION)*, Cambrige, MA, USA, April 2007, pp. 413–422.
43. A. A. Emadzadeh and J. L. Speyer, "On Modeling and Pulse Phase Estimation of X-ray Pulsars," *IEEE Transactions on Signal Processing*, vol. 58, no. 9, pp. 4484–4495, Sept. 2010.

44. ———, "X-ray Pulsar Based Relative Navigation Using Epoch Folding," *IEEE Transactions on Aerospace and Electronic Systems* (in press), 2011.
45. L. Devroye (1986). *Non-Uniform Random Variate Generation.* [Online]. Available: http://cg.scs.carleton.ca/~luc/nonuniformrandomvariates.zip
46. A. Papoulis (2002). *Probability, Random Variables and Stochastic Processes*, 4th ed. New York: McGraw-Hill.
47. A. A. Emadzadeh and J. L. Speyer, "Relative Navigation between Two Spacecraft Using X-ray Pulsars," *IEEE Transactions on Control Systems Technology* (inpress), 2011.
48. A. A. Emadzadeh, A. R. Golshan, and J. L. Speyer, "Consistent Estimation of Pulse Delay for X-ray Pulsar based Relative Navigation," in *Proceedings of the 48th IEEE Conference on Decision and Control*, Shanghai, China, 2009, pp. 1488–1493.
49. A. A. Emadzadeh, J. L. Speyer, and A. R. Golshan, "Asymptotically Efficient Estimation of Pulse Time Delay for X-ray Pulsar Based Relative Navigation," in *AIAA Guidance, Navigation, and Control Conference, Paper AIAA-2009-5974*, Chicago, Illinois, USA, 2009.
50. S. M. Kay, *Fundamentals of Statistical Signal Processing: Estimation Theory*, 2nd ed. Englewood Cliffs, NJ: Prentice Hall, 1993.
51. D. L. Snyder, *Random Point Processes.* New York: Wiley, 1975.
52. N. Ashby and A. R. Golshan, "Minimum Uncertainties in Position and Velocity Determination using x-ray Photons from Millisecond pulsars," in *ION National Technical Meeting (NTM)*, San Diego, CA, USA, 2008, pp. 110–118.
53. P. Panek, "Error Analysis and Bounds in Time Delay Estimation," *IEEE Transactions on Signal Processing*, vol. 55, pp. 3547–3549, Jul. 2007.
54. E. Vonesh and V. M. Chinchilli, *Linear and Nonlinear Models for the Analysis of Repeated Measurements.* New York: Marcel Dekker, 1997.
55. G. Casella and R. L. Berger, *Statistical Inference*, 2nd ed. Belmont, CA: Duxbury, 2001.
56. J. H. Taylor, "Pulsar Timing and Relativistic Gravity," *Philosophical Transactions: Physical Sciences and Engineering*, vol. 341, no. 1660, pp. 117–134, Oct. 1992.
57. G. Jacovitti and G. Scarano, "Discrete Time Techniques for Time Delay Estimation," *IEEE Transactions on Signal Processing*, vol. 41, pp. 525–533.
58. A. Gelb, *Appiled Optimal Estimation.* Cambridge, MA: MIT Press, 1974.
59. J. L. Speyer and W. H. Chung, *Stochastic Processes, Estimation, and Control*, 1st ed. SIAM, 2008.
60. C. T. Chen, *Linear System Theory and Design*, 3rd ed. New York Oxford: Oxford University, 1998.
61. (2010) [Online]. Available: http://webone.novatel.ca/assets/Documents/Papers/LN200.pdf

Index

A
Accelerometer, 87, 88, 97
Accumulated rate, 17
ARGOS, 12
Asymptotic relative efficiency, 57
Asymptotically efficient, 50, 57, 73, 74

B
Brownian motion, 89, 97, 98

C
Consistent (estimator), 50, 57, 62
CPU, 82, 84
Crab, 6, 8, 10, 61
Cramér-Rao lower bound, 2, 35, 37, 111
Cross correlation function, 51

D
Deep Space Network, 4
Delay estimation, 15, 35, 51, 58, 73, 95
Detector, 11, 12, 15, 19, 30, 35
Doppler, 4, 19, 36, 64, 73, 77

E
Empirical rate function, 23, 24, 27, 30, 36, 49, 55, 59, 64, 111
Epoch folding, 1, 13, 22, 23, 25, 27, 49, 59, 64

F
Fisher matrix, 37, 39
Flux, 10, 19, 95
Fourier transform, 58–60

G
Galactic coordinates, 95
Gamma-ray, 6, 8–10
Gaussian, 98
GPS, 4, 5, 12, 14

I
IMU, 14, 87, 97
Inertial measurement unit, 2, 14, 88, 111
Integrated rate, 17, 28
Intensity function, 22, 49
Inverse Fourier transform, 51

K
Kalman filter, 2, 12, 14, 36, 87–89, 92, 95, 111

M
Maximum-likelihood, 2, 73
ML, 36, 59, 73, 78, 79

N
NASA, 9
Non-homogeneous Poisson process, 1, 16
Nonlinear least squares, 55
NRL, 12, 15

O
Observability, 87
Observability matrix, 93
Observable, 93

P
Phase-locked-loop, 12
Photon, 8, 13, 15, 17, 28, 35, 49, 57, 75, 76, 95
Point process, 16

Poisson, 16, 17, 20, 30, 38
Power spectral density, 88
Probability density function, 17, 36
Pulsar profile, 18, 57, 95

R
Radio pulsar, 6, 8, 10, 11

S
solar system barycenter, 14
Spacecraft, 3–5, 10, 11, 13, 35, 58, 73, 87, 111
SSB, 15, 87, 94

T
Time of arrival, 1, 15
TOA, 15–17, 37, 50, 73

U
Unbiased (estimator), 50, 55, 74, 75

W
White noise, 88, 89, 91

X
X-ray pulsar, 1, 3, 6, 7, 11, 13, 18, 87

If you have any concerns about our products,
you can contact us on
ProductSafety@springernature.com

In case Publisher is established outside the EU,
the EU authorized representative is:
Springer Nature Customer Service Center GmbH
Europaplatz 3, 69115 Heidelberg, Germany

Printed by Libri Plureos GmbH
in Hamburg, Germany